Agricultu

DATE DUE

DEC 1 3 1995	
JAN 1 0 1996	
MAR 21 1998	

Food Systems and Agrarian Change

Edited by Frederick H. Buttel, Billie R. DeWalt,
and Per Pinstrup-Andersen

AGRICULTURE AND THE STATE

Growth, Employment, and
Poverty in Developing Countries

EDITED BY

C. Peter Timmer

Cornell University Press

ITHACA AND LONDON

First published 1991 by Cornell University Press.

International Standard Book Number 0–8014–2601–4 (cloth)
International Standard Book Number 0–8014–9911–9 (paper)
Library of Congress Catalog Card Number 90–27706
Printed in the United States of America
Librarians: Library of Congress cataloging information
appears on the last page of the book.

⊗ The paper in this book meets the minimum requirements
of the American National Standard for Information Sciences—
Permanence of Paper for Printed Library Materials, ANSI Z39.48–1984.

Contents

Preface

This volume began in the fall of 1988 when Georg Jakob from Jacobs Suchard, the coffee and chocolate company based in Zurich, stopped in my Cambridge office. The company was about to finish the remodeling of Marbach Castle on Lake Constance as its new executive development center and wanted to inaugurate the new facility with a conference in August 1989. What would be an appropriate topic and how should the conference be organized?

Most academics have a "dream conference" in the back of their minds, and I am no exception. We discussed several topics that related to agricultural development, commodity trade, changing patterns of consumer tastes for food, and the interaction between the public and private sectors in developed and developing economies. I promised to outline in more detail two possible conference topics; he agreed to try them out in Zurich. I put them in the mail the next day.

Nothing happened for two months. Then Georg came to visit again and brought the news that the company had approved a conference on the role of government in agricultural development and would like me to organize it. An ample budget would allow leading scholars to contribute original papers, both to stimulate discussion at the conference and to produce a high-quality book to serve as an academic standard for future Marbach Conferences. The challenge was irresistible. Carol, my wife and long-time editor, and I set out to arrange a series of papers, discussants, and participants for the conference that would meet the expectations of Jacobs Suchard.

Agriculture and the State is the result of that process. Once discussions with potential authors began, the inevitability of the topic became clear. At the end of the 1980s, after a decade of economic restructuring, failed development strategies, and resurgence of market-oriented ideology, the time was ripe for an assessment of what governments should and should not do to stimulate economic growth. Political movements in Eastern Europe, the Soviet Union, China, Latin America, and elsewhere in the developing world challenged the notion that the state always knew best. Especially in the newly emergent democracies in Eastern Europe, the old state was trusted with nothing. An enthusiasm for free markets and unfettered capitalism, warts and all, dominated their economic agendas. For officials in the Reagan, Bush, and Thatcher administrations, and their appointees and converts in the World Bank, the International Monetary Fund, and other donor agencies, the collapse of Marxist approaches and the embrace of markets merely confirmed their decade-long drive to get governments out of economic affairs.

Many of my colleagues in the development profession, while cheered by the victories for personal freedoms and a greater role for market forces, were nonetheless troubled by a failure to read the historical record carefully on the role of government. The dominant theme of the conference became an attempt to do this—to understand, for a handful of important areas involving agricultural development, what defined the *appropriate* role of government intervention. The areas were chosen on the basis of the availability of leading scholars to prepare overview papers and of knowledgeable discussants to criticize and extend the points of view. Chapters on the history of agricultural price policy, on the role of agriculture in export-led growth, on food aid, and on rural development each reflect merely the tip of massive icebergs of ongoing research by the authors on these topics of central importance to agricultural development.

The paper by Just Faaland and Jack Parkinson was commissioned specifically to speak to the broader theme of the role of government in agricultural development. It draws on the authors' decades of experience in Asia and Africa, with a wide variety of governments, in the design, implementation, and evaluation of agricultural policies and projects. Indeed, it was an early comment by Just Faaland, after a particularly frustrating episode—"agricultural development would go better without the government"—that sharpened the central theme of the conference.

The introductory chapter attempts to put the academic debate over these issues into historical perspective. At the level of actual policy debates, the literature is dominated by economists. A more academic approach is taken by political scientists, who have long struggled with the role of government in the development process. Their insights are incorporated in the historical review and in the analysis of several contributors. Political economy is increasingly an applied field of direct use to policy makers; that potential contribution is meant to be reflected throughout.

The final chapter is not a summary of either the papers or the discussions at the conference, although it is influenced by both. Rather, the intent is a more personal assessment of "What have we learned?" Although revisionist glasses now see free trade as the route to success for east and southeast Asian countries that have rapidly growing economies, a more careful historical appraisal demonstrates a complex and varied mix of government and market forces as the engine of growth. Getting the "mix" right, not just "getting prices right," is the key to success, and the mix varies from country to country, from time to time, and from circumstance to circumstance. No single ideological approach provides the right answers any more than does a single technical model.

Agricultural development, a key ingredient for any successful economic development strategy for most poor countries, depends on a pragmatic balancing of economic opportunities, political constraints, and technological possibilities. The goal of this volume is to provide a clearer sense of how history has struck the balance for several of the crucial areas of agricultural development. In particular, the role of the government—what it must do, what it should do, what it should not do, what it cannot do—is identified throughout as the main factor explaining success and failure. We have learned, for better and for worse, that government matters.

This learning process, and the book that attempts to reflect some important lessons from it, would not have been possible without help and guidance from many individuals and institutions. The debt to the managers of Jacobs Suchard is obvious. Without the company's enthusiastic support, both financially and intellectually, this volume would not exist. The chairman of the company, Klaus J. Jacobs, chose the topic, attended the conference, asked some of the toughest questions during the discussions, and hosted a sequence of memorable meals and evenings. Walter Anderau brought some of his key operational personnel to the workshop. They enhanced the learning by authors and discussants about the reality of commodity trade and the food business; I hope we did not shortchange them in the exchange.

It was on Christof Zuber's shoulders, however, that much of the burden fell to make the conference a success. Chris had been in charge of remodeling Marbach Castle in time for the conference and was my day-to-day liaison for working out all the details that must be handled to bring participants from three continents to a successful gathering. All of the conference participants extend a deep vote of thanks to Chris and the superb staff of the Jacobs Suchard Communication and Development Center at Marbach for making the entire experience so rewarding.

Support at Harvard for this endeavor has also been essential. Erin Sands, my assistant at the Harvard Institute for International Development, handled the logistics and communications throughout the entire process with great cheer and enthusiasm, even when the nomadic behavior of authors and the difference in time zones made that difficult. The authors and discussants have indicated that they should thank me for inviting them to such a magnificent setting, which is true enough, but there would be no book without their scholarly attention to these important topics. The discussions themselves would not have been as much fun or as intellectually rewarding without their enthusiastic participation. I do thank them for all their effort and support.

As usual, my deepest thanks must go to my wife, editor, and alter ego. I once noted that Carol seemed not to be bothered by that fine line between improving the wording of a substantive point and changing the substance itself—she crossed it as if she had full diplomatic immunity. Her role at the conference, in the rewriting of chapters, and in the final editing process continues to bear out that observation. Indeed, she must now hold ambassadorial rank—at any rate she threatens to retire!

C. Peter Timmer

Waban, Massachusetts

Contributors

EDWARD CLAY is Director of the Relief and Development Institute in London. Formerly he was a Fellow of the Institute of Development Studies at the University of Sussex, England. He has lived and worked in Bangladesh, India, and Papua New Guinea. He studied for his B.A. degree in economics at Cambridge University and received a D.Phil. degree in economics at the University of Sussex. Recently he has been researching South Asian food policy and food aid.

CRISTINA C. DAVID is an agricultural economist at the International Rice Research Institute. She received her B.A. and M.A. degrees in economics from the University of the Philippines and her Ph.D. degree from Stanford University. Her research and publications have been concerned with agricultural policy and economics of technological change.

GRAHAM DONALDSON is Division Chief of Operations Evaluation at the World Bank. Before joining the Bank in 1970, he worked for ministries of agriculture in Australia, Britain, and Canada and taught at Wye College in Britain. He holds degrees in agriculture from the University of Sydney and a Ph.D. degree from the University of London.

JUST FAALAND is the Director General of the International Food Policy Research Institute (IFPRI). In the postwar era he was an economist in the Organization for European Economic Cooperation (OEEC) in Paris in the years of the Marshall Plan. He then joined the

Chr. Michelsen Institute in Bergen, Norway, where he built up the institute's Department of Social Science and Development. He has been an economic adviser to several governments of developing countries and to international organizations and is a member of the United Nations Committee for Development Planning. He was president of the OECD Development Centre in the early 1980s. Dr. Faaland did his undergraduate and advanced studies at the University of Oslo and Oxford University.

WALTER P. FALCON is Director of the Food Research Institute and Helen C. Farnsworth Professor of International Agricultural Policy at Stanford University. He is Chairman of the Board of Trustees of the International Rice Research Institute and a member of the Board of Trustees of Winrock International. Dr. Falcon, who received his B.A. degree from Iowa State University and his Ph.D. degree from Harvard University, received the Distinguished Achievement Citation from Iowa State University in 1989. His current research interests include American agriculture policy, food policy in Indonesia, and the world food economy.

RAYMOND F. HOPKINS is Chair of the Department of Political Science at Swarthmore College. He is also Director of the Food Policy Program and has served as coordinator of the international relations concentration and Director of the Center for Social and Policy Studies. He has worked on issues of politics and agriculture since the 1970s and has been a Guggenheim and Rockefeller Fellow. He has taught as a visiting professor at the University of Pennsylvania, Columbia University, and Princeton University.

JAMES P. HOUCK is Professor of Agriculture and Applied Economics at the University of Minnesota, where he has served since 1965. His professional interests center on agricultural trade, trade policy, and prices. He received his B.S. and M.S. degrees from Pennsylvania State University and his Ph.D. degree from the University of Minnesota. He was editor of the *American Journal of Agricultural Economics* in 1981–83.

PETER H. LINDERT is Director of the Agricultural History Center and Professor of Economics at the University of California–Davis. His research interests span international economics and modern economic history. He received his A.B. degree in public and international affairs from Princeton University and his Ph.D. degree in economics from Cornell University.

JACK PARKINSON is Professor Emeritus of the University of Nottingham and has had wide experience in economic planning. During World War II, he was in the Prime Minister's Statistical Branch in the United Kingdom and worked in Europe with the Marshall Plan. In Northern Ireland he was a member of the Economic Council. He has been a consultant and researcher in Asia, particularly Pakistan and Bangladesh, and in Africa and has contributed to several books on economic development.

C. PETER TIMMER is Thomas D. Cabot Professor of Development Studies, At Large, Harvard University. He has been a resident adviser in Indonesia and has worked on food and agricultural policy in China, Japan, the Philippines, Sri Lanka, and, recently, Vietnam. He received his A.B., M.A., and Ph.D. degrees in economics from Harvard University and has taught at Stanford and Cornell universities. He is a Faculty Fellow of the Harvard Institute for International Development and teaches economic history, the economics of the world food system, and business and government relations.

ALBERTO VALDÉS serves as a senior researcher in the Latin American Department of the World Bank. Formerly he was Program Director of the International Trade and Food Security Program at IFPRI, Washington, D.C. Dr. Valdés graduated from Catholic University of Chile and received his Ph.D. degree in economics from the London School of Economics. While at IFPRI, he conducted research on food security issues in less-developed countries, agricultural trade relations between industrial and developing countries, and the linkage between trade and the exchange rate regime and agriculture.

Agriculture and the State

I

The Role of the State in Agricultural Development

C. Peter Timmer

Two grand themes have long dominated the debate over the role of the state in a society's economic affairs: how to stimulate rapid economic growth and how to reduce the level or consequences of poverty. Economists often characterize this debate as a matter of finding the optimal trade-off between efficiency and equity (thus implicitly assuming that efficiency leads to growth). Policy analysts tend to stress the importance of being on the frontier in this trade-off between growth and alleviation of poverty in the first place and thus are more likely to recognize the potential of state intervention to stimulate growth from starting points in the interior of the frontier. Political scientists, however, see either characterization as too narrow, leaving out their broader concerns over the reciprocal issue of the nature of the state itself and its contribution to growth and alleviation of poverty. Because the level and structure of development and the degree of poverty obviously influence both the potential for state intervention and the type of response, this broader view of the debate incorporates a wide range of social, political, and economic dynamics that are not easily included in economists' growth models.

This volume attempts to grapple with several of these broader themes without becoming lost in either vague generalities or a unique vision of the path to utopia. The emphasis is on the contribution of agricultural development to growth of the entire economy and to the alleviation of poverty. Achieving these objectives normally requires implementation of multiple policy instruments, and the various chapters focus on several: price and trade policy, generation of employment, rural development programs, and food aid. No reader familiar with the analytical and historical record of the postwar era, especially the decade of the 1980s, will be surprised to learn that no single program or even sectoral policy approach is sufficient to meet *either* goal of stimulating growth or alleviating poverty. Close integration of agricultural policy with overall macroeconomic and trade policy seems to be essential for rapid economic growth, and rapid growth seems to be essential to sustained reductions in poverty.

The complexity of designing and implementing such an integrated economic development strategy, even one that emphasizes market coordination, will clearly challenge even the most competent and growth-oriented state. Where the political reality is more diffuse and economic leadership less influential, the process of meeting the goals of both growth and alleviation of poverty will be much more difficult. But it is precisely that political reality and the policy environment it creates that explain most of the diversity of development experience in the past several decades. Any attempt to discuss the role of the state in agricultural development in terms of pragmatic policy choices must address both this diversity and the political realities that lie behind it. This broad political economy perspective is a major unifying theme of the chapters that follow.

The Evolution of the Debate

Modern economic policy analysis was developed to address the issue of state intervention into agriculture. The debate over the Corn Laws in early nineteenth-century Britain pitted Ricardo's increasingly sharp and focused microeconomic models against Malthus's vague but realistic concerns for dynamic macroeconomic and general-equilibrium effects. This tension between the concretely measurable and the vaguely important surfaces repeatedly in the efforts of economists to construct empirically based and verifiable models of the impact of government actions on an economy. The problem is not just whether macroeconomic models have solid microeconomic foundations but whether comparative statics models of equilibrium positions have any real value in achieving an understanding of the dynamic paths for which continuous nonequilibrium outcomes may be a better description of reality. Understanding these dynamic paths is crucial in any attempt to trace the consequences for economic growth of government-induced changes in the agricultural sector.

The foundation of the neoclassical approach to this question is the volume by T. W. Schultz, *The Economic Organization of Agriculture,* published in 1953. Scholars who are familiar only with later writings by Schultz on rational farmers and the importance of price incentives will be surprised to find a careful analysis of imperfections in factor markets in United States agriculture coupled with a pervasive concern for the powerful negative welfare effects generated by instability in commodity prices and incomes of American farm families. Only the government is capable of dealing with the macroeconomic forces that were the basic cause of this instability, and Schultz provides a carefully analyzed set of interventions that would lessen the social impact of commodity price instability, including income transfers during periods of depressed economic activity. But no satisfactory approach to dealing with instability within agriculture is found:

> The instability of farm prices is an important economic problem. It is, however, exceedingly difficult to organize the economy so that farm prices will be on the one hand both flexible and free and on the other relatively stable. Farm price supports and efforts to control agricultural production by acreage allotments, marketing quotas, and related public measures are not satisfactory. Diversion operations, subsidized exports, and efforts to shelter the domestic market from foreign competition are also unsatisfactory. (Schultz, 1953, p. 365)

Schultz does quote approvingly the work of D. Gale Johnson on forward pricing for agriculture as an approach to improving the efficiency of resource allocation in the agricultural sector.

Schultz's preference for dealing with fundamental causes rather than symptoms shows up in 1964 in his most influential work, *Transforming Traditional Agriculture.* Farm incomes are low in developing countries not because peasants are irrational or lack knowledge of how to farm efficiently with the resources at their disposal but because they lack the technology that would generate higher incomes. The demise in the late 1960s of community action programs and the emphasis of donors and policy makers on research and development of new agricultural technology can be traced fairly directly to Schultz's influence. The subsequent failure of the Green Revolution to transform the agricultural sectors of poor countries at anything like its advertised potential is explained in 1978, in *Distortions of Agricultural Incentives,* as the consequence of government policies that prevent prices in world markets from reaching farmers in developing countries. Pressures from urban consumers, the need for tax revenues, and instability in world markets served to reinforce policy

makers' biases against a positive role for agriculture in the development process. In this increasingly Chicago-esque view of the world,
government intervention into agriculture is almost always wrong-
headed and harmful. Markets, with all their failures, work better to
serve the interests of agriculture and the rest of the economy.

The debate over the role of government in agricultural development has, then, become increasingly focused on the tensions between
market failures and government failures. A series of contributions in
the 1980s has begun to clarify the issues analytically. Binswanger and
Rosenzweig (1986) attempted to draw on fundamental characteristics
of agricultural production processes and household behavior to explain patterns of organization for different crops and the potential
for government intervention to improve the efficiency of resource allocations. Several of these characteristics, especially the importance
of imperfect and asymmetric information, led Stiglitz to examine the
prevalence and efficiency of linked markets in agriculture. In many
cases, the underlying market failures that prevented unconstrained
Pareto optimal outcomes were not susceptible to improvements by
cost-effective government intervention, especially land tenure arrangements and rural credit markets (Stiglitz, 1987, 1989). Yet the
failure of risk markets may offer important opportunities for governments to stimulate rural investments and "learning by doing" by appropriate pricing policies for inputs and outputs. Most important,
Stiglitz emphasizes the empirical nature of the debate:

> Simple prescriptions have one obvious advantage: they enable econo-
> mists to make policy judgments with virtually no knowledge of the
> country in question. But the flaws in the simplistic approach make it all
> the more important to accumulate detailed information on most of the
> developing world. The first step in any systematic analysis of agri-
> culture policies is therefore to describe as accurately as possible the con-
> sequences of each policy. This requires a model of the economy
> concerned—and a model appropriate for one country may not be for
> another. Recent work has clarified some of the essential ingredients of
> these models: wage-setting policies in the urban sector, the nature of
> rural-urban migration, and the organization of the rural sector—labor,
> land, and credit markets. . . .
>
> In any analysis of agricultural policies, the hardest part is to incorpo-
> rate political economy considerations—to decide what are to be taken
> as political constraints. (Stiglitz, 1987, p. 54)

Precisely because it is the core of the problem, the political economy of agricultural policy has begun to attract the attention of econ-

omists and political scientists. Bates, Srinivasan, Anderson and Hayami, and Varshney have used political economy models of rent seeking, rational choice, and organizational costs to explain pervasive patterns of government intervention into the agricultural sectors at different levels of development. Two of these patterns, the "development paradox" and the "anti-trade bias," are analyzed in considerable historical detail by Lindert in Chapter 2. What is still missing in the literature, however, is much consideration of how actual policy settings shift and evolve and how policy analysts can use sudden shifts in economic or political environments as "windows of opportunity" to pursue more effective policies for the economy, including the agricultural sector. The recent widespread experience with structural adjustment may well provide the necessary insights to generalize about this issue.[1]

Government Intervention in Agriculture

Agreement on the role of government in agricultural development spans a fairly wide continuum, from provision of basic law and order, which even supporters of Hayek or Mises would acknowledge, to direct control of daily field operations for nearly all agricultural commodities, which nonrevisionist Maoist and Stalinist believers still claim is necessary to mobilize resources for state-directed central plans. A convenient dividing point, however, is where mainstream neoclassical economists split on the *principle* of state intervention. Agreement in principle is widespread that governments should intervene to provide public goods and to correct other significant market failures. Actual interventions require careful analysis of costs and benefits, including the potential of government failure to be worse for growth prospects than the market failure ostensibly being corrected. There is, of course, substantial scope for controversy in designing and evaluating these analyses, and much of the modern empirical record in development economics concerns these issues. But it is a controversy almost entirely within the standard neoclassical paradigm.[2]

[1]See Perkins and Roemer (1991) for a collection of papers that discuss the experience of the Harvard Institute for International Development with reform of economic systems.

[2]For example, volume 1 of the recent *Handbook of Development Economics* contains sixteen chapters, nine of which might be classified as empirically oriented and directly relevant to policy. All of the latter use neoclassical methodology. See Chenery and Srinivasan (1988).

On the other side of the divide is a set of issues upon which economists do not agree, even in principle: whether state intervention is potentially helpful or harmful. Many of these issues involve complex and often vague trade-offs between economists' measures of efficiency and the distribution of welfare consequences from the development process. The debate, then, is frequently between incommensurate values and approaches. Such debates inevitably are resolved by political means, and the revival of political economy as an analytical field has begun to provide rigor and structure to the academic debate (Bardhan, 1988). But economics has much to contribute if the issues are clearly stated so that key trade-offs can be identified and quantified. A major purpose of this book is to stimulate the contribution of analytical economics to this complex but extraordinarily important debate on the role of government in agricultural development.

Government Interventions: Agreement in Principle

It is widely agreed that governments should provide public goods and correct important market failures. The debate is over levels of funding, appropriate institutional organization, and relative roles for public agencies and private firms and households. Since Pigou, this debate has formed an integral core of analytical economics, and the modern field of public economics has evolved to deal directly with these issues. The major topics in agricultural development are government support and organization for research, extension, irrigation, and the rural marketing infrastructure.[3]

Agricultural Research. The public good aspects of agricultural research have been recognized by governments for centuries, well before economists provided a formal analytical rationale for widespread public support for improving agricultural technology.[4] Optimal incentives to private firms to invest in the discovery of new technology require that the new income streams generated be appropriable to a

---

[3]See Chapter 10 for a discussion of how the nature of the state affects this debate.

[4]The early history in Great Britain is analyzed by Lerner (1989), who finds that agricultural research in the seventeenth and eighteenth centuries provided significant stimulus to basic science. Such externalities and the perceived inability of individual "improvers" fully to capture the value of their innovations were used as arguments in support of more public support for development (and dissemination) of agricultural technology. These issues are also discussed in Timmer (1969).

significant degree by the firm incurring the costs of research. Although hybrid seeds with secret inbred lines, patented chemical formulas, or specific brand-name farm implements meet this criteria and consequently are activities of the private sector in developed countries, most technology for food grains, livestock, and inputs falls outside this category. The inability of private firms to capture more than a tiny fraction of the increased financial flows made possible by innovations in these commodities means that there will be little research activity by them unless it is directly funded by the public sector.

Scale economies in modern research also militate against optimal levels of research being carried out by the private sector in developing countries. Indeed, agricultural producers now conduct only a trivial amount of research on new seeds and techniques in all countries, in contrast to earlier in the century, when most productivity-enhancing innovations arose on the farm and were spread by word-of-mouth. Modern science and technology have nearly eliminated the role of the "farmer improver" championed by Arthur Young in Great Britain and Charles Warren in the United States.

Partly because modern science has stimulated a technological revolution in the way agricultural research is done (and the subsequent productivity of the innovations) and partly because the public goods often receive inadequate budgetary support, the returns to public sector investment in agricultural research are typically very high. Research by Griliches (1958), Evenson and Kislev (1975), and others document internal rates of return on such investments at double and more the social opportunity cost of capital. David, in her comments on Donaldson's analysis of the World Bank's experience with rural development projects in Chapter 6, cites these high rates of return as the basis for criticizing the seriously inadequate support for agricultural research in such projects. The absolute necessity for new technologies to generate higher income streams for traditional farmers was stressed by T. W. Schultz (1964) in the volume that provided the analytical foundations for an entirely new approach to agricultural development. Clearly, only in countries with the capacity to fund and conduct the agricultural research that yields these new technologies can agricultural development take place at a rapid enough pace for the sector to play a broader role in stimulating the entire development process. Building this capacity to provide the financial and scientific resources required deserves very high priority.

Agricultural Extension. Although "research and extension" are often said in the same breath as though they were one concept, in fact the "extension" part of the story is much more controversial than the research part. To be sure, there are empirical counterparts to the high economic returns from agricultural research (although sometimes the high payoff to extension is a result of lumping research and extension costs together in the analysis), but there are also many documented failures and doubts about the capacity of extension services to deliver anything of value to farmers. Because research results are virtually useless unless adopted by farmers, nearly all public research programs are designed to have outreach components to extend their results to the farm level. The controversy over extension programs arises not because of the need for such components as a public sector activity—the public good aspect is clear enough—but over program design itself. The widely hailed "training and visit" approach, or "T&V," supported by the World Bank, has demonstrated high returns in some environments. In others, critics contend, the initials stand for "talk and vanish." The crucial problem is establishing a link between the constraints that farmers perceive on their own farms and the research agenda of agricultural experiment stations and scientists. Extension programs that have been designed as one-way flows from "scientists who know" to "farmers who don't" usually fail in settings where low productivity stems from complex interactions across crops and practices in diverse agronomic environments— rather than single bottlenecks for a particular crop in a widely uniform agro-climatic zone. It was relatively easy in 1978 for an extension agent on Java to explain that the recently released rice variety IR-36 was resistant to brown planthoppers in irrigated rice paddies but much more difficult to explain improved cultural practices for soybeans in upland fields that were subject to many pests and diseases. Without a full understanding of the problems faced by soybean farmers, agricultural scientists were likely to be solving the wrong problems, thus leaving extension agents with nothing to extend.

Wide cultural and educational gaps between scientists, government extension agents, and farmers seem to be major reasons for continuing problems in establishing effective two-way channels of communication. No simple and direct mechanisms exist to rectify this problem—certainly not organizational changes in government extension services. Basic investments in literacy programs for farmers, rural communications networks, and public commitment to the value of agriculture in the society will be needed over a substantial period of time to close the cultural and educational gaps. No routine cost-benefit appraisal of such investments will capture their contribution to a more effective research and extension effort even though this external value may be sufficiently important to society to justify substantial expansion in budgetary support to these basic investments.

Irrigation Investments. The appropriate mix between public and private investments in irrigation is a lively issue. Although there is no intrinsic reason why large-scale dams and delivery canals cannot be a private sector activity, private firms in developing countries seem unable to manage the coordination, design, finance, and execution of such projects. Only tubewell irrigation and low-lift pumps from rivers and canals are primarily a private sector activity in developing countries (and the developed world as well).

Because large-scale irrigation schemes are almost always a public sector activity, various policy issues arise. At one level there is concern about the impact of many irrigation facilities on public health and the environment. Diversion of water from natural flows inevitably has some consequences for the environment, including the potential for downstream salinity problems, the creation of breeding grounds for such public health hazards as schistosomes and malarial mosquitoes, and depletion of underground aquifers that may be important water sources some distance away.[5] In principle, these externalities should be included in the evaluation of costs and benefits to the public sector irrigation project. Indeed, the presence of such important externalities is a major reason why large-scale irrigation projects should be a public sector activity. But the actual track record of incorporating environmental and public health costs into the design and evaluation of irrigation projects is dismal indeed, whether the projects were funded by external donors such as the World Bank or came directly from the country's own budget.

At this level of concern, Hayami and Kikuchi (1978) have demonstrated the existence of a powerful "irrigation cycle" of public sector investments. When grain prices are high in world markets, signaling scarcity to both private and government buyers, countries increase their budgetary commitments to expanding agricultural output, especially through investments in irrigation facilities to grow rice and wheat. Multilateral and bilateral donors follow the same short-run

---

[5]For a useful overview of the public health dimensions of agricultural development in Africa, see Ohse (1988).

economic logic in evaluating their investments. The result is a surge of investment in irrigation shortly after world markets signal a grain shortage. Of course, grain production from these investments does not appear on domestic and international markets for nearly a decade because of the long lags between identification of a need for an irrigation system and its construction and operation. As these supplies flow onto markets more or less in synchronous fashion, world prices are depressed. The low prices then cause governments and donors to reject further investments in irrigation, thus setting the stage for the next iteration of the cycle. The failure, of course, is the short time horizon for price expectations on the part of governments and donors. Although this behavior is somewhat understandable on the part of governments that must pay the current market price for imports, the failure of donors to take a longer-term perspective on the effect of commodity prices on their investment decisions is distressing and perplexing.

At the grass-roots level, several issues regarding water pricing become important. Simple concerns for allocative efficiency suggest that farmers should pay a fee related to the volume of water they use and the economic cost of delivering or replacing it. Bureaucratic efficiency in operation and maintenance of irrigation facilities by public sector employees suggests that the fees paid by farmers should also be connected to the timeliness of water deliveries and the quality of water services. And a broader concern for the integrity of public sector budgets suggests that full cost recovery from beneficiaries of public sector irrigation projects is needed to provide the resources for continued investments.

Virtually none of these private charges are actually paid. Most countries have provided irrigation water free (or at modest fixed charges) to farmers, with both investment costs and operations and maintenance charges paid out of the budget of the central government (or sometimes by state or regional authorities). As a consequence, actual budgets for operations and maintenance are usually seriously inadequate even for an efficient bureaucracy. More important, there are no incentives to perform the operations and maintenance activities in a timely and effective manner, and most irrigation systems need complete rehabilitation well before their economic and technical designs would indicate. Investment costs are then spread much more thinly than is necessary, thus slowing the expansion of agricultural output.

Government failures such as these have led to calls to privatize a greater proportion of development-related investments, including those in the irrigation sector. But it is not easy to see how the private sector can overcome the barriers to its effective involvement. Two compromises are possible. One draws on the best features of the two sectors to produce the right volume of irrigation projects, in the right places, and with appropriate concern for externalities; design, construction, and operation are efficiently managed by the private sector. The other compromise combines the worst of the two sectors: public sector funding subsidizes private investors to ignore cost-recovery issues, to scrimp on operations and maintenance programs, and to evade any responsibilities for negative externalities. The risk of the second approach is sufficiently large in most countries that efforts to improve the capacity of the public sector to design, finance, and manage irrigation programs are probably better investments than trying to find new institutional arrangements that privatize most of these decisions.

Marketing Infrastructure. After two decades of declining support for public investment in the rural infrastructure that contributes to efficient marketing of agricultural commodities, the direct and indirect contributions of this infrastructure to rural growth and reductions in poverty are again being recognized.[6] The decline was caused to some extent by tighter budgets in most developing countries, but sharp criticisms in the 1970s of the supposed failure of investments in rural infrastructure to reach the "poorest of the poor" also reoriented priorities among donors toward meeting basic needs through rural development programs. For example, as Donaldson points out in Chapter 6, the World Bank committed itself to double its lending for poverty alleviation, primarily through rural development projects in which more than half the direct beneficiaries were to be below the poverty line.

The importance of an efficient marketing system in raising the productivity of an impoverished rural population is now recognized (and is stressed by Donaldson in Chapter 6), but public sector investments to provide the physical infrastructure for such a system were badly neglected in the 1970s and early 1980s. This neglect shows

[6]See in particular recent research in Bangladesh carried out by the International Food Policy Research Institute (IFPRI) and the Bangladesh Institute of Development Studies reported in Ahmed and Hossain (1990) and Kumar (1988).

clearly in the empirical record of agricultural exports reported by Valdés in Chapter 3, and on changes in labor productivity in the agricultural sector in Asia reported by Timmer in Chapter 5. And yet substantial controversy remains over how many public resources should be devoted to rural infrastructure and what share of the needed investments should come from the private sector.

An examination of the key areas of concern reveals the reasons for the controversy. For a rural marketing system to work efficiently, an entire set of interlinked components must be in place and mesh relatively smoothly. These components include farm-to-market roads, regional highways, railways, trucks, and rolling stock; communications networks involving telephones, radios, and information-gathering capacity; reliable supplies of electricity for lighting, to operate office equipment, and to power rural industries; market centers and wholesale terminals with convenient access to both transport facilities and financial intermediaries; and a set of accepted grades and standards for traded commodities that permit reliable "arm's-length" contracts to be written and enforced at low cost.[7]

Each individual component of a well-functioning marketing system has a major role for private sector involvement, perhaps even to the exclusion of any necessary public role. Certainly trucking companies, warehouse operators, and rural and regional banks can be entirely in the private sector, and, indeed, the empirical record suggests that they should be if reasonable efficiency standards are to be maintained. But a marketing system is more than the sum of these private firms, partly because the links that connect these firms—the roads, railways, telephone networks, and so on—have important public good dimensions or problems of coordination that markets alone have a difficult time solving. A further part of the story, however, involves substantial economies of scale and externalities in the construction and operation of the marketing system itself. No private firm can hope to capture the full economic benefits accruing to an efficient marketing system, even when the firm's own investment is a crucial component needed to make the system work.

The existence of these externalities and systemwide scale economies that cannot be appropriated by individual private firms creates an important role for the public sector in guaranteeing that the basic rural infrastructure is in place and operates efficiently. Clearly, direct

[7]Chapter 4 of Timmer, Falcon, and Pearson (1983) discusses at greater length these components of an efficient marketing system and analyzes government policies that stimulate its development.

public investment, ownership, and operation are neither necessary nor even desirable in many contexts, and regulation, indicative plans, and appropriate investment incentives may well be sufficient. But equally clearly, a direct public role may be needed in many circumstances, especially in the building of roads, railroads, and communications networks. These types of capital-intensive investments, however, came under fire in the 1970s as failing to help the poor. Expanded funding for them now, especially in the context of universally tighter public budgets, requires a clear rationale.

Investment in infrastructure has two important economic payoffs. Rural infrastructure, in the form of irrigation and drainage works, roads, ports and waterways, communications, electricity, and market facilities, provides the base on which an efficient rural economy is built. Much of the investment needed to provide this base comes from the public sector, even when the private sector is playing the predominant role in agricultural production and marketing. Without this public investment, rural infrastructure is seriously deficient in stimulating greater production of crops and livestock. Investment by the private sector is also less profitable in the absence of adequate rural infrastructure, thus further reducing rural dynamism. Public sector investment in rural areas has a "crowding in" rather than a "crowding out" effect on private investment, and for this reason the main role of investments in infrastructure is this longer-run stimulation of agricultural production, which has important positive effects on rural employment and income distribution.

A second role needs to be stressed as well. Investments in infrastructure can directly generate substantial rural employment, and this potential has not been lost on planners seeking both long-run employment creation and short-run work programs to alleviate rural poverty or even famine conditions. Food-for-work and employment guarantee schemes almost always are designed to build rural infrastructure using low-cost or unemployed workers. Large-scale irrigation and road construction projects offer the potential to employ vast numbers of unskilled rural laborers if project designers are sensitive to employment issues in the choice of technique and are willing to address the managerial problems that arise from labor-intensive techniques in construction.

The progressive commercialization of agriculture—as more productive inputs are purchased and a greater share of output is marketed, made possible by the development of an efficient marketing system—is a major stimulus to agricultural productivity and creates substantial employment in the agriculture-related industries. In modern economies far more workers are engaged in agribusiness than in farming itself. In the less-developed agricultural economies, such nonfarm but agriculturally-linked employment is not quite so important. Even so, the single most important sector of the industrial labor force is usually in agricultural processing. Employment in rice or wheat milling, jute mills, cotton spinning and weaving, and cigarette manufacture is often the main source of organized factory jobs. When small-scale traders, food wholesalers, retailers, and peddlers are included, the volume of indirect employment begins to rival direct employment on farms. Many of the workers are the same, or at least live in the same household. Half of the income for farm households on Java now comes from off-farm labor. Not all of the jobs are in large- or small-scale agribusiness, of course, but most are linked via well-functioning commodity and factor markets to the health of the rural economy (and the strength of the urban construction industry).

Relatively few policy instruments are available to stimulate efficient investment in the agribusiness sector. Parastatal and state-owned enterprises have a poor record of commercial viability. Their employment record may be adequate in terms of numbers of workers, but labor productivity—value added per worker—tends to be very low. More efficient firms and more productive workers are the result of a competitive private sector, and stimulating the development of such firms is now a high priority of most countries. Because so many impediments to the private sector have existed historically, especially in agribusiness and marketing, policy reforms that remove barriers to private sector participation are an important first step. But stimulating private investment while creating a competitive market structure is a delicate task, not one in which most governments have any real experience. Policies that restrict licenses to a limited number of firms in order to guarantee market share might induce investment, but they produce an oligopolistic market structure. By contrast, an aggressive competition policy is likely to scare off private investors, especially domestic entrepreneurs, at least initially.

It is fairly apparent that simply "getting prices right" in the agricultural and marketing sectors does not of itself induce the necessary private investments or competitive market structure. Inappropriate price policies are like other barriers to participation by the private sector; removing them might be necessary but not sufficient, in the absence of other institutional and legal reforms, to guarantee greater private investment. Economists are woefully ignorant of the basic causes of the "animal spirits" that motivate private investors, but the need for a competitive market structure is compelling to the profession. Businessmen are happy to explain what they need to make a profit; a government guarantee of that profit would then lead them to invest. Striking the right balance between the two perspectives will take pragmatic experimentation with alternative policies.

Government Interventions: Widespread Disagreement

The above discussion amply demonstrates that there is no agreement on policy even if the role of the government is agreed upon in principle. When the role itself is up for debate, the controversies are sharper and deeper. Economic models provide no clear-cut guidelines, and historical experience, ideology, and short-run political dynamics provide most of the ammunition for policy debates. The issues that dominate this agenda are important, however, both to the dynamic efficiency of the economy and to its potential for rapidly alleviating poverty. Four issues regarding the role of the government are particularly relevant: reform of land tenure relationships, farmer organizations such as cooperatives, the operation of marketing boards, and pricing interventions seasonally and with respect to world prices. In all these areas a tempting generalization might be that what looks good in theory works badly in practice and what looks bad in theory is widely practiced and sometimes has good results. In all four cases well-intentioned policies often have perverse effects on intended beneficiaries, especially on impoverished rural inhabitants, because efforts to improve income distribution—at the heart of these controversial activities—frequently carry unforeseen losses in efficiency that hurt the long-run prospects of the poor.

Land Tenure. "Land to the tiller" campaigns and other efforts to redistribute land in a more equal fashion usually claim both efficiency and equity gains. Indeed, at the level of principle a land reform is arguably the shortcut to problems of poverty. A country in which agricultural land is owned and operated predominantly by smallholders whose labor productivity in the production of food crops is high should have a relatively equal distribution of income and adequate access to food. But the poor have basic needs other than food, farms might be too small to support large and growing families, and labor productivity could be threatened by population growth and inadequate development of new technology.

The important question for policy is how to break into the nexus between equity and efficiency. Should a country strive for better "initial conditions" by undertaking land reform, concentrate on rapid improvements in labor productivity and real wages, or attempt direct programs to alleviate poverty to improve the distribution of basic goods and services? There are trade-offs among the three possibilities, if for no other reason than that the government's budget has many claimants. But the trade-offs run much deeper, into the basic economic and political mechanisms that dictate how a country's economy produces and distributes output. Land reform is a political exercise with surprisingly few solid economic underpinnings. Although this is no doubt a controversial statement, it stems from experience with nonrevolutionary land reforms since the 1950s, especially as reviewed in the debate over the desirability of a land reform in the Philippines after Marcos. Although nearly all economic analysts support some form of land reform, they do so primarily for reasons of political dynamics or distributional equity. The recent "neo-neo-classical" literature on interlinked markets has significantly undermined the earlier Marshallian view that only owner-operators could use land efficiently.[8] Without strong efficiency gains, the economic case for land reform becomes much weaker, especially if substantial disruption occurs to established patterns of input supply and output marketing. Most policy makers grant the desirability of more equal distribution of land but will want to know if progress on improving income distribution can be made in other dimensions.

Farmer Organizations. A powerful mystique surrounds farmer cooperatives, which have become the most important form of rural organization now that China has disbanded its communes. By banding together, the theory holds, farmers can share the risks of crop failures, achieve economies of scale in purchasing inputs and selling output, and attain equal or superior market power in dealing with monopolistic middlemen who perennially exploit individual small farmers. In some circumstances, the theory clearly works. Danish cooperatives helped transform a backward agricultural economy into a competitive export sector in the latter half of the nineteenth century, farm cooperatives in the United States were instrumental in the adop-

[8]The seminal works in this field were Cheung (1969) and Stiglitz (1974).

tion of legislation that regulated private railroad tariffs, and Japanese cooperatives have provided secure access to farm inputs and the political clout to protect agricultural interests from foreign competition. In each case, the cooperatives improved farm incomes and probably increased overall productive efficiency so that there were gains to society at large as well.

These success stories have provided a rationale for governments in developing countries to foster their own versions of rural cooperatives. Nearly always the creatures of government initiative, centrally directed cooperatives have a dismal track record of improving farmer welfare. Part of the reason is that many governments saw cooperatives as a vehicle for helping farmers in ways that substituted for more expensive or politically difficult assistance, such as research on new technology, investments in rural infrastructure, and adequate price incentives for farm output. Simply put, cooperatives cannot substitute for any of these essential ingredients in agricultural development and improvements in rural welfare.

But even in the context of supportive policies for agriculture, cooperatives have had an extremely difficult time establishing a credible and financially viable role in rural economies. The problems are primarily generic and not specific to country or decade. Perhaps the fundamental problem is the total neglect of the high transactions costs involved in organizing and managing a geographically scattered clientele in an environment in which communications networks are poor or nonexistent. Unless the benefits to organization are significantly larger than these transaction costs, farmers will inevitably drop out. The second faulty assumption is that marketing costs are high because middlemen make excess profits through monopoly power. If that were true, cooperatives could capture these profits for their members by providing the same marketing services directly. In fact, very few marketing cooperatives have been able to provide a similar range and quality of services as those of their private sector competitors at any cost, much less a lower one. Again, farmers drop out to get access to input supply credits, more timely deliveries, and cash payment for crop sales.

Risk sharing via cooperative membership has two problems. First, most cooperatives draw their members from the same agro-ecological zone, so weather risk is highly covariant. A drought for one member usually means a drought for all. Second, to the extent that production returns are shared through cooperative rules, significant incentive effects will sharply reduce the overall effort and quality of managerial input in farming. These incentive effects eventually led to the demise of Chinese communes, especially in their early versions, when output was to be shared by vast numbers of members, often 50,000 or more.

A final problem for rural cooperatives has been the difficulty in obtaining and retaining the skilled management needed to run a complex organization. Especially when the cooperatives have been set up by governments, diverse tasks in input and output marketing, handling of financial assets and repayments of loans, and providing technical information to farmer members call for a broad range of managerial talent. Such talent usually has a high opportunity cost in these societies, and, especially if trained in urban universities or institutes, managers have a strong preference for the amenities of city life. Only when cooperatives have trained their own leadership are they able to retain capable managers, and even then the technical skills of such leaders may be seriously deficient. Perhaps more to the point, home-grown cooperative leadership usually conflicts directly with the desire of the central government to control the entire cooperative movement from the top. By divorcing cooperative leadership and management from grass-roots members, however, such a top-down approach makes it virtually impossible to internalize the high transactions costs of organizing effective cooperatives in the first place. It is no wonder that so few government-sponsored cooperatives have succeeded in improving either productivity or income distribution.

Marketing Boards. An alternative to cooperatives as a solution to the problem of scattered farmers with relatively small volumes of produce for sale is for the state to take over the marketing directly, either for output alone or for the key inputs as well. Colonial governments found marketing boards a highly effective device for collecting exportable commodities, stabilizing their prices, and extracting substantial tax revenues in the process. Rather than overthrowing these colonial vestiges, most newly independent countries found their continued role to be essential for revenue purposes and to provide political patronage.[9]

The very rationale for a state-controlled marketing board—to tax export commodities and stabilize their internal prices— requires that it maintain a monopoly over export sales. Otherwise, private traders

[9]Both the history of marketing boards and the political imperative for their continued existence are documented by Bates (1981).

could avoid paying the tax and profit whenever the divergence between world prices and domestic prices for the commodity favored private exports. The same logic does not, however, require that the marketing board maintain a monopoly in domestic marketing between farmers and the point of export, and private traders could easily be permitted to compete for this trade. Two factors have tended to prevent this approach. First, an ideology that supports state control of exports also tends to support the abolition of middlemen because of their reputation for price gouging and monopolistic exploitation. Second, a relatively competitive private marketing sector almost always has lower costs and better information than state marketing boards. In the absence of large subsidies or legal monopoly powers, these state boards are usually unable to procure supplies in domestic markets and must rely on the private sector to supply their export orders. This dependence makes most managers of marketing boards very uncomfortable, especially if supplies in the hands of the private sector might find alternative outlets across porous borders or in black markets. Whenever a primary objective of operating a marketing board is to control the flow of the commodity, there is an inevitable tendency to squeeze out the private sector.

With the power to control the flow of commodity comes the power to extract tax revenues for the public sector and rents for the parties in control. When the sector that produces agricultural commodities is the main source of net value added in an economy, as in much of Africa, the very existence of the state depends on this capacity to extract revenues. As Bates (1981) has shown, legitimate taxation to fund the basic infrastructure of the state is easily turned into the foundation of political empires through payoffs in the form of patronage for supporters. Given the inherent necessity to generate tax revenues, however, the crucial question is whether any other mechanisms exist that would be less susceptible to capture by politicians or would at least be more accountable to the efficiency concerns that provide the basis for rapid economic growth.

It is fairly easy to see why many economists increasingly answer this question by proposing the abolition of marketing boards in favor of broader-based value-added taxes and incentive-neutral, revenue-producing trade taxes. In most cases, they argue, any market failures that result from allowing the private sector to handle all agricultural marketing, both domestically and internationally, will have smaller negative consequences for economic growth than are generated from the current forms of marketing organization and state control. Any need to stabilize producer prices can be implemented as a separate policy with independent instruments, as Valdés points out in Chapter 3. With a vigorous pro-competition policy for the domestic marketing sector, possibly implemented by a state-supported marketing board with a charter to compete with the private sector, farmers will no longer be at the mercy of either a public or private monopoly on marketing services. A vigorous supply response is expected when heavy tax burdens and high marketing margins are removed, and economic growth will be stimulated.[10]

Although the performance of most marketing boards in developing countries has no doubt had pernicious effects on the growth process and a private marketing sector will perform better in the long run, the issue remains of how to move from one organizational form to the other. Once again, economists seem to have a clear vision of where the economy should be but remarkably few levers to nudge it gradually in the right direction. Dismantling state controls in favor of market processes is, however, a leading agenda item—from Poland and the rest of the Eastern bloc to all of the Third World. Economists would have more to offer during the process of change if their models focused less on comparisons of static equilibria and more on political economy dynamics during the disequilibrium of structural and institutional change. Fortunately, the intuition of many economists who work on these problems is better than their models, and the profession seems to be moving rapidly in directions that will improve our capacity to understand the dynamics of economic systems reform.

Price Interventions. In a sense, marketing boards are merely one of many vehicles a government can use to intervene in domestic price formation, and both the level and stability of export prices fall in their domain. Also, marketing boards have often been set up for domestic food crops as well as export crops, and stabilization of consumer prices has been a key part of their mandate. But the issue now is broader: should the government be intervening at all in the formation of agricultural prices, especially staple food prices? That is, does "getting prices right" mean getting government out of price formation and allowing free trade instead?

From the perspective of economics, only two possible reasons for an interventionist food price policy might be defended: the interven-

---

[10]The leading exponents of this approach have been Bauer (1976) and Lal (1985). Disenchantment with state marketing boards is now, however, very widespread, and most donors actively seek to dismantle them as part of any policy-based lending.

tions could improve the efficiency of the economy and thereby speed economic growth, or they could improve income distribution and raise the welfare of the poor.[11] No persuasive case for keeping food prices substantially above or below their long-run opportunity costs in world markets has been demonstrated either theoretically or historically. The efficiency losses from either bias in domestic pricing eventually become a substantial burden even in rich countries, and the budgetary costs of continuing subsidies have direct consequences for the rate of economic growth.[12]

The failure to make a case for sustained price biases relative to long-run opportunity costs in world markets is not, however, an argument for free trade in food grains. Policies that stabilize food prices along the long-run trend in world market prices have the potential to contribute simultaneously to both economic efficiency (and growth) and improved income distribution through better nutritional welfare of the poor. This argument is not new to the economics profession, but the analytical case for food price stability has never been put in a sufficiently dynamic and macroeconomic context for the benefits to appear large relative to the costs of stabilization programs.[13]

A consensus has emerged in the economics profession that the welfare gains from price stabilization, although usually positive on theoretical grounds, are empirically not very important relative to the costs governments must incur to stabilize prices. Consequently, little rationale exists for governments to attempt to stabilize prices of food grains. But this conclusion, and its obvious implications for government policy, is sharply at variance with what governments actually do. When an objective such as price stability is universally held to be politically important, there is often an underlying economic rationale that eludes narrow analytical models. In such circumstances, a different analytical approach is needed if economists are to contribute to the policy debate.

Three key innovations in such an approach, two microeconomic and one macroeconomic, offer the potential for very different empirical conclusions. The first change in microeconomic approach is to consider the farmer as an investor rather than the manager of a static stock of assets and a flow of variable inputs. The model of farmer as

[11]Strictly speaking, a neoclassical economist would defend pricing interventions to redistribute real incomes only if nonprice redistributions, such as lump sum transfers or asset redistributions, were impossible for bureaucratic or political reasons. Such is often the case, however.
[12]These lessons are explained in greater detail in Timmer (1986).
[13]For a more extensive treatment of this topic, see Timmer (1989).

manager is the basis of nearly all theoretical and empirical assessments of risks from price and yield instability, but this model excludes important elements in farm decision making that are strongly influenced by these risks, especially expectations and patterns of investment in physical and human capital. Transforming the problem into one of dynamic portfolio investment decision making enormously complicates the analysis of risk, even when restricted to farm-level issues.[14]

The second change in microeconomic approach allows the consumer to value price stability for reasons other than implied changes in consumer surplus. In particular, when food costs are a large component of the budget, any changes in food prices will imply substantial reallocation of budgetary resources among substitutes and complements. These changes involve decision-making efforts with significant transactions costs, and these costs should be included as a negative element in the consumer's welfare function. The size of the transactions costs and their importance to the consumer's overall welfare are a function of income levels, and the poor suffer more significant welfare losses from frequent price changes than richer consumers who are under less pressure to reallocate their budgets.

Macroeconomic benefits from stabilization of food prices stem from the inelastic demand for food and from the importance of food prices in determining the real wage in developing countries. Frequent changes in either nominal or real wages affect both the demand for and supply of labor, with potential spillovers into levels of labor-intensive exports, government tax revenues, inflation, and the real exchange rate. Unfortunately, tracing the macroeconomic ramifications of price instability is even more complicated than adapting microeconomic models of farmers and consumers because general-equilibrium analysis is needed with dynamic investment functions that are conditioned by stability-sensitive expectations. But incorporating these dynamic factors into both the micro and macro analyses offers the opportunity to examine the impact of price-stabilization policies on agricultural development and economic growth. The static, micro-based models do not address these issues; they are incapable of assessing the consequences for the economy of the price-stabilization policies that are widely implemented—consequences that policy makers actually worry about.

---

[14]For a relatively simple example of how complex the analysis becomes, see Rosenzweig and Binswanger (1989).

No formal model capable of addressing these wider concerns is offered here, but the likely ingredients of such a model can be readily identified. Too much price instability for staple food prices will cause displaced investments in physical capital at the farm level, in the marketing sector, and in the industrial sector; substitution of consumption and leisure for savings and work; biases in investments in human capital for the farm agent and intergenerationally in children; substantial transactions costs for consumers when reallocating budgets as prices change; the welfare losses from a psychic sense of food security (voters in rich and poor countries alike place a substantial economic price on this factor); and the feedback from this sense of security to a stable political economy, which reinforces investors' willingness to undertake long-term (and hence risky) commitments.[15]

Demand for price stability cannot readily be expressed in markets, especially in developing countries. For this reason, the popularity of price stability is usually treated by economists as a political issue, not as one for economic analysis. Such a narrow analytical perspective misses the inherent market failure underlying the transfer of society's desire for price stability from a nonexistent economic market to the political arena, where the demand for a public good can be expressed and met. The resulting increase in social welfare is very much an economic phenomenon, comparable to the welfare generated by the consumption of "real" goods and services.

When elements such as these are factored into the analysis, the benefits from stabilizing the prices of basic foodstuffs are likely to be considerably larger than those reflected in the models that have been used so far to analyze relative costs and benefits of price-stabilization programs.[16] When properly managed, food price stabilization programs have the potential to improve economic efficiency and thereby speed economic growth. Although little is known empirically about the size of the dynamic and macroeconomic benefits of stability, they cannot be ignored in the theoretical or empirical evaluation of such programs. For example, the pervasive, indeed universal, tendency of Asian governments to stabilize their domestic rice prices in the face of unstable world market prices for rice suggests that the benefits may be very large. The rapid economic growth in many of these Asian countries indicates that the impact of efficiency losses and budgetary

---

[15]Each of these components is discussed at length in Timmer (1989).

[16]Stiglitz (1989) recognizes that price-stabilization schemes, if properly designed, may capture some of these more dynamic benefits.

costs on growth cannot be too large, at least if the price-stabilization program is well designed and implemented.

Themes and Variations

The most obvious conclusion from the above discussion is that the appropriate role of the state in agricultural development is an empirical question which requires sophisticated policy analysis to determine in each particular setting. Every country is different; its history, institutions, and economic structure differ from those of even its closest neighbors. But policy analysts do not start in a vacuum, required to reinvent the wheel to make any progress. General themes and lessons do exist, analytical methodologies are quite robust, and human behavior has a commonality of purpose that permits the borrowing of insights from one society to illuminate the problems in another.

This introductory chapter has attempted to follow several of these general themes within the framework of a loosely defined political economy methodology that is based primarily on neoclassical economics. No apologies are needed for the analytical framework so long as the problems to be analyzed are tossed up by the process of governments interacting with their agricultural sectors in search of paths to higher standards of living for the entire society, rather than by the methodologies themselves. In some sense, that search also provides a roadmap to the chapters in this volume. Partly by design and partly by the serendipity of individual scholars pursuing their own sense of priorities, the following chapters trace a sequence of analytical stories that reflect the inherent tension between unique circumstances and the desire to seek general explanations for policy advice.

The story begins with the lessons of history. In Chapter 2, Lindert uses a spare model of political economy to explain two apparently perplexing and yet historically robust patterns of state intervention in agricultural pricing: the "development paradox" in which rich countries with few farmers transfer large sums to the agricultural sector through price interventions, whereas poor countries with many farmers tax their agricultural sectors via the same instruments; and the "anti-trade bias" in which most countries protect their import-competing agricultural sectors and tax their agricultural exports. In Chapter 3, Valdés shows the extent to which currently developing countries exhibit both patterns and documents the severity of the discrimination against agricultural producers, which is a result of both direct and indirect economic policies. The key role of macroeconomic policies, especially management of the exchange rate, is stressed if countries are to adopt a strategy of economic growth based on expansion of agricultural exports.

Chapters 2 and 3 are primarily concerned about the efficiency consequences of state intervention, especially through price policies (directly and indirectly). Government concern for poverty alleviation is introduced in Chapter 5, where Timmer focuses on efforts to generate employment as a vehicle for raising the incomes of the poor. The contrasting experiences of South and Southeast Asia between 1960 and the 1980s reveal that measures to raise labor productivity in the agricultural sector can be successful only in the context of well-functioning factor markets, urban-rural links, and a competitive industrial sector. Rural poverty remains endemic until agricultural labor productivity rises significantly, and both an effective agricultural *and* an industrial development strategy are needed to raise real wages for unskilled workers.

Concern for poverty alleviation in rural areas is explored directly in Chapter 6. Donaldson's review of the World Bank's experience with rural development programs reinforces the need for appropriate technologies to raise productivity in less-favored rural environments and the critical role of trade and an efficient marketing sector in this effort. The need for locally effective leadership in the context of vast institutional changes facing many rural communities appears as a major bottleneck to speedy replication of the success stories highlighted by Donaldson, and outside resources, both financial and human, may be crucial to starting and maintaining the process.

One potentially important outside resource is food aid. Especially in settings where fairly immediate improvements in food consumption are necessary to alleviate poverty, food aid may be the only viable option for countries with little access to commercial food supplies on world markets. In Chapter 8, Clay demonstrates that this food aid potential is real but that the historical record of its effectiveness is quite poor. Improving the fit between supplies of food aid, still highly subject to the whims of major donors and to opportunity costs in world markets, and need for food aid in circumstances that will not discriminate against local producers, remains the most pressing policy issue for the food aid community. Clay argues persuasively that no resolution of this issue is possible unless it is incorporated into a broader food policy analysis that includes macroeconomic policy as well. Falcon points out in his discussion, however, that countries with good food policies in the first place seldom need significant quantities of food aid.

The capacity of any government to conduct such a broad food policy analysis and to implement the results depends fundamentally on the nature of the state itself. As Faaland and Parkinson argue in Chapter 10, most governments in developing countries are "soft" in Myrdal's sense and so are predominantly concerned about maintaining power rather than maintaining the long-run vision needed for policies that stimulate economic growth. The capacity to stay in power often depends primarily on support of the army, the government bureaucracy, and organized urban workers. Such a power base rules out favorable treatment for the agricultural sector, with the ultimate loss of its essential contribution to the development process. Faaland and Parkinson are fairly pessimistic about redressing this fundamental bias, for the reasons discussed above in the context of reforming economic systems. The pressures for reform are building in those societies with poor economic performance, however. The concluding chapter offers some observations on how the development profession might be better prepared for opportunities as they arise.

References

Ahmed, Raisuddin, and Mahabub Hossain. 1990. *Developmental Impact of Rural Infrastructure in Bangladesh*. IFPRI Research Report no. 83. Washington, D.C.: International Food Policy Research Institute; October.

Anderson, Kym, and Yujiro Hayami, with associates. 1986. *The Political Economy of Agricultural Protection: East Asia in International Perspective*. London: Allen and Unwin.

Bardhan, Pranab. 1988. "Alternative Approaches to Development Economics." In Hollis Chenery and T. N. Srinivasan, eds., *Handbook of Development Economics*, vol. l, pp. 39–71. Amsterdam: North-Holland.

Bates, Robert H. 1981. *Markets and States in Tropical Africa: The Political Basis of Agricultural Policies*. Berkeley: University of California Press.

Bauer, P. T. 1976. *Dissent on Development*, rev. ed. Cambridge: Harvard University Press.

Binswanger, Hans P., and Mark R. Rosenzweig. 1986. "Behavioral and Material Determinants of Production Relations in Agriculture." *Journal of Development Studies* 22: 503–39.

Chenery, Hollis, and T. N. Srinivasan, eds. 1988. *Handbook of Development Economics.* Vol. 1. Amsterdam: North-Holland.

Cheung, S. N. 1969. *The Theory of Share Tenancy.* Chicago: University of Chicago Press.

Evenson, Robert E., and Yoav Kislev. 1975. "Investment in Agricultural Research and Extension: An International Survey." *Economic Development and Cultural Change* 23, no. 3: 507–21.

Griliches, Zvi. 1958. "Research Costs and Social Returns: Hybrid Corn and Related Innovations." *Journal of Political Economy* 66 (October): 419–31.

Hayami, Yujiro, and Masao Kikuchi. 1978. "Investment Inducement to Public Infrastructure: Irrigation in the Philippines." *Review of Economics and Statistics* 60 (February): 70–77.

Kumar, Shubh K. 1988. *Rural Infrastructure in Bangladesh: Effects on Food Consumption and Nutrition of the Population.* Washington, D.C.: International Food Policy Research Institute.

Lal, Deepak. 1985. *The Poverty of Development Economics.* Cambridge: Harvard University Press.

Lerner, Joshua. 1989. "Science and Agricultural Development: Quantitative Evidence from England, 1660–1780." Cambridge: Department of Economics, Harvard University. Typescript.

Ohse, Takateru. 1988. "Epidemiology of Hunger in Africa." In David E. Bell and Michael R. Reich, eds., *Health, Nutrition, and Economic Crises,* pp. 223–40. Dover, Mass.: Auburn House.

Perkins, Dwight H., and Michael Roemer, eds. 1991. *Reforming Economic Systems.* Cambridge: Harvard University Press.

Pigou, A. C. 1920. *The Economics of Welfare.* 1st ed. London: Macmillan.

Rosenzweig, Mark R., and Hans P. Binswanger. 1989. "Wealth, Weather Risk and the Composition and Profitability of Agricultural Investments." Minneapolis: Department of Economics, University of Minnesota. Typescript.

Schultz, Theodore W. 1953. *The Economic Organization of Agriculture.* New York: McGraw-Hill.

———. 1964. *Transforming Traditional Agriculture.* New Haven: Yale University Press.

———, ed. 1978. *Distortions of Agricultural Incentives.* Bloomington: Indiana University Press.

Srinivasan, T. N. 1985. "Neoclassical Political Economy, the State and Economic Development." *Asian Development Review* 3, no. 2: 38–58.

Stiglitz, Joseph E. 1974. "Incentives and Risk Sharing in Sharecropping." *Review of Economic Studies* (April): 219–55.

———. 1987. "Some Theoretical Aspects of Agricultural Policies." *World Bank Research Observer* 2 (January): 43–60.

———. 1989. "Market Failures and Development." *American Economic Review* 79 (May): 197–203.

Timmer, C. Peter. 1969. "The Turnip, the New Husbandry, and the English Agricultural Revolution." *Quarterly Journal of Economics* 83 (August): 375–95.

———. 1986. *Getting Prices Right: The Scope and Limits of Agricultural Price Policy.* Ithaca: Cornell University Press.

———. 1989. "Food Price Policy: The Rationale for Government Intervention." *Food Policy* 14 (February): 17–27.

Timmer, C. Peter, Walter P. Falcon, and Scott R. Pearson. 1983. *Food Policy Analysis.* Baltimore: Johns Hopkins University Press for the World Bank.

Varshney, Ashutosh. 1989. "Political Power and Food Prices: Food Price Policy in India since 1965." Ph.D. diss., Department of Political Science, Massachusetts Institute of Technology, Cambridge.

2

Historical Patterns of Agricultural Policy

Peter H. Lindert

The governments of today's industrial countries distort their economies by protecting their farmers. Governments of developing countries, by taxing their farmers heavily, distort their economics even
more. The explanation for this apparent tendency is just emerging.
Since comparative analysis of agricultural pricing began in the
1970s, analysts using postwar data have revealed two patterns: *the
developmental pattern*—the more advanced the nation, the more its
government favors agriculture—and *the anti-trade pattern*—governments tend to tax exportable-good agriculture and protect import-
competing agriculture.

The developmental pattern implies that successfully developing nations will drift toward subsidizing agriculture, although the record
does not point to an obvious ending point in the most advanced nations. The anti-trade bias implies that policy will dampen comparative advantages the world over, cutting dependence on and gains
from international trade in agricultural products. Both patterns seem
to conflict directly with economists' models of efficient resource allocation. But research on this issue is relatively new and cuts across
several disciplines, not just economics, extending well beyond traditional agricultural policy analysis.[1]

---

[1]*Agricultural policy* in this chapter means all government policies affecting the real income of persons in the agricultural sector, including policies not explicitly aimed at agriculture. The breadth of this definition invites the inference that all indirect effects of policy
are known to policy makers in advance—and to scholars in hindsight. That cannot be true.

The history and theory of today's international patterns in agricultural policy have just begun to receive systematic attention.[2] Did these patterns prevail anytime before World War II, or are they postwar phenomena? Why should there be consistent patterns at all among nations so varied in their politics and geography?

This chapter attempts to provide both a history and a theory of the developmental and anti-trade patterns in agricultural policy. Both patterns existed before World War II but have been hidden by two clouds: English exceptionalism and the Great Depression of the 1930s. Consideration of basic economic influences can help explain these patterns, but these have been hidden by two other clouds: the belief that the history of each nation is unique and its policies defy global influences, and the willingness to use popular arguments for agricultural protection as if they explained the observed policy patterns.

This chapter is organized into three main sections. The first explains the primary measure of comparing agricultural price policies—the nominal protection coefficient (NPC) that compares domestic prices with world market prices for the same commodity—and notes the biases in the approach. The biases understate the patterns, however. Better data and methodology would sharpen the policy contrasts between more- and less-developed nations and between agricultural-exporting and agricultural-importing nations.

Yet for expositional convenience, I argue as though policy makers and those pressuring them tended to see most of the influences that scholars can see. I discuss those few main policies covered by the conveniently available data, omitting many others. My implication that the ones covered had the greatest effect cannot be defended in detail.

[2]This is not to say, of course, that no contributions have been made to understanding the history and theory of the comparative policy patterns. Parts of the history have been surveyed by sources cited here. For an early comparative survey of the history of protectionism over two centuries, see McCalla (1969). Explanations of the recent comparative patterns were offered by the authors documenting them. Yet the historical literature has not addressed the recent patterns, and the explanations are still preliminary and ad hoc, as elaborated below.

Several contributions provide a base to understanding the history and theory of the comparative policy patterns. Most well-known empirical tendencies in economics are based on long historical experience or on common folklore and are sanctified with quotes showing that they were foreseen by Adam Smith back in 1776. The developmental pattern of agricultural policy, however, emerged only in the 1970s on the basis of postwar data analyzed by Little, Scitovsky, and Scott (1970); Balassa and associates (1971); Johnson (1973); Schultz (1978); Lutz and Scandizzo (1980); Bale and Lutz (1981); and Binswanger and Scandizzo (1983). The World Bank, the FAO, and other international agencies have played a leading role as providers of data and sponsors and publicists for much of the relevant research, for example, World Bank (1982, 1986) and the forthcoming book edited by Schiff and Valdés. A pathbreaking pair of books by Anderson and Hayami (1986) and Tyers and Anderson (forthcoming) has extended the pattern to other commodities and back into the 1950s for fifteen leading countries.

The modern patterns described by the NPC measure are presented as a base to follow government treatment of the agricultural sector backward in time. The patterns of today can be traced back to the 1860s, the Great Depression notwithstanding. Different patterns stand out in the history of the less industrial, more impoverished world before the 1860s. Central government (usually the throne) often squeezed agriculture, especially in food crises, when it sought to force delivery of affordable food to the cities. That is, there was an urban bias similar to that observed in the developing countries today. Agriculture fared better at the hands of governments dominated by landed elites. Ironically, this group included the government of the first industrial nation.

The historical survey clears the way for a deeper and more complex theory for explaining the patterns of agricultural policy, which is developed in the second section of the chapter. Traditional and politically popular rationales offered for agricultural protection—for example, the goal of food security, concern over the instability of the agricultural sector, sympathy for farmers as poor people, and political nostalgia for the farm sector—fail to explain who taxes agriculture and who supports it, even though these popular explanations certainly are elements in the story.

This failure stimulates the development of a frugal pressure-group model that underlines two basic reasons why agricultural policy should shift from taxing to subsidizing agriculture. First, the shrinkage of the agricultural population raises the per capita subsidy and lowers the per capita cost of a given real transfer. Second, the relative size and strength of the political alliances defending agriculture do not shrink nearly as fast as the sector itself, because the barriers to political participation by farmers and their closest allies decline faster than the barriers to participation by others.[3] Apparent exceptions, such as England during the Corn Laws (1660–1846), are shown to validate the general model by being as peculiar in their political structure as in their policies.

In addition, the model brings out two ways in which the switch in agricultural policy should have been more dramatic over the course of economic development than changes in the policy toward any other major sector. First, the decline in population share, mentioned above, is historically more dramatic for agriculture than for other sectors. Second, own-price elasticities of supply and (especially) demand

[3]A similar point was made by Mancur Olson (1985).

are lower for agricultural products than for other sectors of comparable size and trade orientation, thus reducing the efficiency losses of
government price interventions.[4]

Several supplementary arguments make the pressure-group model
more robust. Historical evidence shows a secular rise in the sensitivity of farm-sector income to changes in agricultural prices. The ability of the government to control these price fluctuations creates
strong pressures to do so. In addition, the government's own demand
for revenue helps explain the anti-trade bias of policy, especially in
newly independent nations.

The third section of the chapter provides a statistical model of the
pressure-group theory and tests it with an extensive data set using
observations from around 1980. A nested set of hypotheses is formulated using the arguments developed in the theoretical section, and
regression analysis is used to test the competing explanations for the
observed policy patterns. As expected, the popular explanations for
price interventions do not survive this more rigorous testing, but several of the theoretically specified variables do. Much of the diversity
remains unexplained at a deeper level, however, because even the
highly significant effects of sector size and income variables may still
be related to "development" itself. The concluding section speculates
on what forces may be operating at this deeper level.

Patterns of Agricultural Price Interventions

To compare policies and reveal patterns over dozens of countries,
one needs a readily available and robust statistic that describes the
policy or measures its impact. The simplest approach, and the one
used in most of the recent empirical literature, is a comparison of
domestic with border prices using the nominal protection coefficient
(NPC). This section analyzes the biases involved in using this measure and concludes that any patterns revealed by the coefficients will
show up even more clearly with more sophisticated measures. It then
uses the NPC methodology to examine the modern cross-section pat-

[4]Lower price elasticities mean lower net deadweight costs of marginal redistribution
through government. Gardner (1987, 1988) has shown that governments in industrialized
countries tend to cut deadweight costs further by choosing among farm supports so as to
minimize the relevant elasticities and costs. Government intervention is not resisted as
strongly when the main redistribution takes place through agricultural markets, both in
the early settings in which the debate was usually over how much to tax agriculture and in
the industrialized nations debating how much to subsidize it.

tern of agricultural price interventions. Finally, the historical roots of this pattern are examined for both modern industrial countries and the Third World.

Methodology and Measurement

To compare policies over dozens of countries, analysts carrying out recent comparative studies have had to do without relevant details and concentrate on a single statistic available from every country. That statistic, the nominal protection coefficient, is the ratio of the domestic producer price to the border price, or world price, of the same agricultural product, sometimes with adjustments for overvaluation of the domestic currency. If the NPC is above unity, farmers are said to be protected at the rate NPC − 1. If it falls below unity, they are said to be taxed at the rate 1 − NPC.

Nominal protection coefficients fall far short of quantifying the impact of all government policy on the relative income of the agricultural sector. They omit any taxes or subsidies that do not affect the farm-gate price of the product. Two micro-level measures that avoid part of this defect are the effective rate of protection (ERP), widely used in the 1960s and 1970s, and the producer subsidy equivalent (PSE), developed by economists at the Organization for Economic Cooperation and Development and the U.S. Department of Agriculture.[5] The ERP captures the protective (or taxing) effect of unit subsidies (or taxes) on internationally traded inputs into the relevant output sector. The PSE may miss some of the subsidies and taxes from trade policies relating to farm inputs, but it has the advantage of quantifying some of those other payments and taxes that do not affect prices of traded goods. In principle, the PSE can be refined into a fairly comprehensive measure of the effects of all policies on the rate of producer surplus.

Two other key shortcomings of the NPC also afflict the currently available ERP and PSE measures. All three are measures of effects on nominal incomes within the (farm) enterprise, and they ignore any effects of government policies on the cost of living and thus the real value of those nominal incomes. Obviously, taxes and tariffs (or subsidies) on consumer products imply a direct markdown (or markup)

[5]See OECD (1987), USDA (1988). A measure similar to the ERP is the "total" nominal protection rate. See Schiff and Valdés (forthcoming). It includes the traded-good price effects of the ERP plus an exchange-rate adjustment, which is made by some, but not all, ERP studies.

in the real incomes of farmers. The other shortcoming is the failure to include systemic (for example, general-equilibrium) influences of all policies on the market prices of inputs and outputs. Trade policy, for example, affects wage rates, interest rates, rents and exchange rates, and other real prices that are taken as given in the NPC, ERP, and PSE measures.

Despite the shortcomings of NPCs as measures of governmental impact on the real purchasing power of farm income recipients, they are likely to show the correct direction of agricultural distortions in the postwar era, with some understatement of the net taxation of agricultural products in developing countries.[6] When we turn to historical patterns, the NPCs are likely to be freer of bias, because they capture the effects of an import tariff or an export tax, the policy tools that dominated before 1930.

Cross-Section Patterns in the Modern Era

Nominal protection coefficients from around 1980, shown in Table 2.1 and Figure 2.1, suggest both the trade bias and the developmental pattern of agricultural policy. To illustrate the trade bias, three important products—wheat, rice, and beef—have been separated from the export crop category. NPCs for exportable crops are less than one, meaning that the domestic producer price is below the world price, usually because of official restrictions on exports. Producers of importables, by contrast, enjoy higher NPCs for any given level of GNP per capita, as illustrated in the cases of wheat, rice, and beef. Even without developing a model of the policy process, one can think of a proximate explanation for the trade bias: governments find it easier to protect producers of importables and tax producers

---

[6]For two reasons, NPCs understate the contrast between the taxation of agriculture in developing countries and its protection in industrialized countries. First, the elaborate calculation of PSEs still leaves the same commodities in the taxed or protected camps in the countries studied (see USDA, 1988). Second, the ERP studies have shown that the developing countries, though nominally taxing agriculture, are heavily protecting their manufacturing sectors, imposing much greater taxes on agriculture through their purchases of farm inputs and consumer goods. Estimates from six developing countries in the 1960s show that their NPCs understate the absolute and especially the relative taxation of agriculture (see Balassa and associates, 1971, tables 3.2 and 3.3). In a study of eighteen developing countries during the 1975–84 period, industrial protection and overvaluation of the domestic currency were shown to reduce the effective protection of agriculture by 11 percent of export crops and 20 percent of importable crops (see Krueger, Schiff, and Valdés, 1988, pp. 262–63). In addition, the available estimates captured only indirect commodity taxation, missing the frequent tendency of direct taxes to fall more heavily on agriculture in developing countries.

of exportables because both policies can generate tax revenue and do not require subsidies.

The development pattern of agricultural policy also emerges in Table 2.1 and Figures 2.1a–c. For each product, there was a rough upward slope relating net protection to GNP per capita around 1980. In this international cross section, the slope is due almost entirely to the discrete jump in NPCs across the income gap separating developing from developed countries. The raw correlation between NPC and GNP per capita is insignificant among developing countries. It is also lacking among developed countries, partly because two high-income nations, Canada and the United States, were relatively unprotective agricultural-exporting nations in 1980.[7] But there does seem to be a difference between the two groups of countries. No industrial country has resisted the pressure to subsidize its producers of importable goods, though the North American and Australasian governments have kept fairly neutral toward their exportable agricultural products.

A food-commodity pattern also exists, but is hardly mentioned by the literature. Among the fifteen industrial nations studied by Anderson and Hayami, the general ranking by NPC for 1980 is

Milk > beef > sugar $\geq$ grains > chicken, eggs, and pork

The ranking for 1980 had some glaring exceptions: eggs received heavy price protection in Australia and New Zealand, and beef was much more heavily protected in the European Community than elsewhere. Yet there is a pattern of sorts, running from dairy products down to pork, chicken, and eggs. Fruits and vegetables are difficult to categorize, but they were generally less protected than grains (Anderson and Hayami, 1986; Tyers and Anderson, forthcoming).

This general ranking of protection and the exceptions to it need explanation even if they are unique to the postwar economy. But if

[7]When raw correlations are replaced with descriptive regressions, there is a hint that protection peaks and begins to retreat when the share of agriculture reaches 6 to 8 percent of the labor force and 4 percent of GNP in 1980. So say the regression runs by Honma and Hayami (1986, pp. 43–47) on pooled data for fifteen nations, 1955–80. From 1980 to 1986, however, protection surged in Canada and the United States, possibly yielding a monotonically positive slope of protection with respect to GNP per capita. See USDA (1988) and Carter (1988).

The NPC estimates themselves seem to contain measurement errors that significantly reduce the goodness of fit of any regression in which NPC is the dependent variable. I have found several cases of multiple estimates of the NPC for the same country and commodity around 1980. The range of discrepancy is worrisome, though not so great as to overthrow the developmental or anti-trade patterns of policy.

Table 2.1. Nominal protection coefficients for agricultural products, ca. 1980

		Net protection coefficients					
		Three products, in nations not exporting them			Export crops	Export crop	Source[b]
Country	1983 GNP per capita[a]	Wheat	Rice	Beef			
Bangladesh	130	0.63	0.69				B and S
Mali	160				0.48	cotton	B and S
Malawi	210				0.65	groundnuts	B and S
Tanzania	240				0.39	coffee	B and S
India	260	0.80	0.65				B and S
Kenya	340	1.13	1.30		0.74	coffee	B and S
Pakistan	390	0.76	0.68		0.61	cotton	B and S
Sudan	400	0.95			0.48	cotton	B and S
Senegal	440				0.46	groundnuts	B and S
Egypt	700	0.76	0.76		0.44	cotton	B and S
Sub-Saharan Africa	700	1.09	0.34		0.73	all export crops	WDR '86
Côte d'Ivoire	710		0.97		0.39	cocoa	B and S
Philippines	760		0.73				B and S
Cameroon	820		0.50		0.32	cocoa	B and S
Thailand	820				0.39	rubber	B and S
Turkey	1,240	0.94	1.50		0.54	tobacco	B and S
Tunisia	1,290	0.99					B and S
Colombia	1,430		0.92	0.92			B and S
Brazil	1,880		0.66	0.60	0.43	coffee	B and S
Korea	2,010	1.33	2.56	1.67			A and H

Country	GNP per capita					Commodity	Source
Argentina	2,070				0.64	wheat	B and S
Portugal	2,230	0.71	0.60	1.49	0.66	olives	B and S
Yugoslavia	2,570	0.76		0.76			B and S
Italy	6,400	1.41		2.27			A and H
New Zealand	7,730	0.65					A and H
United Kingdom	9,200	1.08		1.82			A and H
Netherlands	9,890	1.44		1.95			A and H
Japan	10,120	2.92	3.61	2.00			A and H
France	10,500	1.11		1.88			A and H
West Germany	11,430	1.20		1.90			A and H
Australia	11,490				0.99	sugar	A and H
Denmark	11,570	1.07		1.73			A and H
Canada	12,310				0.90	wheat	A and H
Sweden	12,470	1.06		2.14			A and H
United States	14,110				0.82	wheat	A and H
Switzerland	16,290	2.77		2.49			A and H

Note: The three sources take different approaches to adjusting for disequilibrium exchange rates. Both the World Bank and the Binswanger-Scandizzo study tried to adjust their nominal protection coefficients for disequilibrium exchange rates; Anderson and Hayami did not. This difference should not mar comparability much, because the industrial countries covered by Anderson and Hayami had minimal exchange controls.

[a]The figures for GNP per capita in 1983 are from World Bank, *World Development Report, 1985* (New York: Oxford University Press for the World Bank, 1985).

[b]The sources for the nominal protection coefficients are as follows:

A and H: Kym Anderson and Yujiro Hayami, with associates, *The Political Economy of Agricultural Protection: East Asia in International Perspective* (London: Allen and Unwin, 1986).

B and S: Hans P. Binswanger and Pasquale L. Scandizzo, "Patterns of Agricultural Protection," World Bank Discussion Paper ARU 15 (Washington, D.C.: World Bank, 1983).

WDR '86: World Bank, *World Development Report, 1986* (New York: Oxford University Press for the World Bank, 1986).

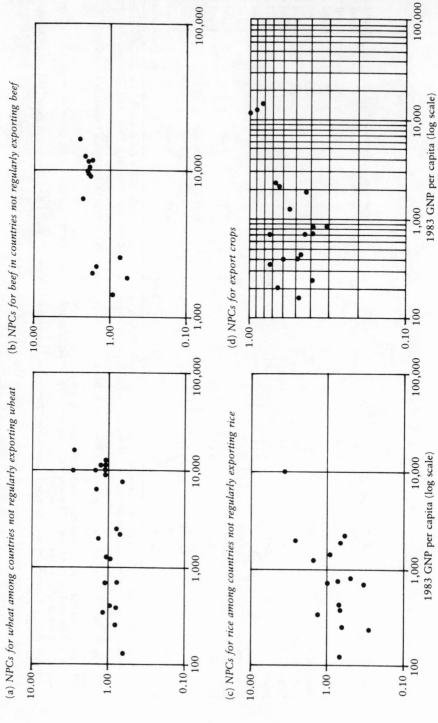

Figure 2.1. Nominal protection coefficients for agricultural products, ca. 1980

(a) NPCs for wheat among countries not regularly exporting wheat

(b) NPCs for beef in countries not regularly exporting beef

(c) NPCs for rice among countries not regularly exporting rice

(d) NPCs for export crops

1983 GNP per capita (log scale)

the patterns have deeper historical roots, the nature of the explanation must also be deeper. Before attempting to build a political economy model of agricultural policy, it is necessary to see just how pervasive the patterns to be explained are in historical terms.

Historical Patterns

The historical questions to be addressed parallel those asked by modern price policy analysts. Do both the development pattern and the anti-trade bias in policies affecting agriculture appear in the historical record? Did the developed countries of today tax agriculture more heavily in the past when their economies were poorer and more agricultural? Did they favor import-competing producers of agricultural products over producers of exportables? What were the earlier policies of governments of countries that are still less developed today? Even this fairly rough survey of the historical experiences of these countries finds significant evidence of longer-term historical patterns. These patterns match those of modern experience, but there are several equally puzzling and important exceptions, which complicate the task of any model designed to explain them.

Industrial Countries. Examination of the historical record of the United States and the industrialized countries of Europe and East Asia reveals an uneven trend toward agricultural protectionism. Rates were highest in the depressed 1930s, followed by those around 1972, just before the inflationary oil shocks.

The history of Japan clearly conforms to the developmental pattern. Tax and trade policies shifted from taxing to subsidizing agriculture over the century since the Meiji Restoration. Direct taxation discriminated against agriculture under the Tokugawa shogunate up to 1868. The Meiji Restoration continued to tax agriculture more heavily than industry, but the net tax rate dwindled until the tax rates on agriculture and nonagriculture were roughly equal—and low—in the 1930s. (Ranis, 1959; Hayami, 1988, pp. 18–50).

Japan began protecting domestic rice farmers against import competition as early as 1904. That was the year Japan first became a steady net importer of rice, needed revenues for the war against Russia, and was no longer bound by the free-trade clauses of earlier treaties signed with the great powers. Rice protection and imperial self-sufficiency grew until rice farmers received a net protection of 84

Figure 2.2. Nominal rates of protection against rice imports into Japan, 1900–82

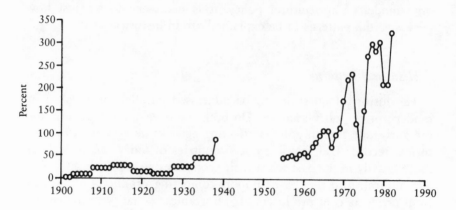

Source: Data from Kym Anderson and Yujiro Hayami, with associates, *The Political Economy of Agricultural Protection: East Asia in International Perspective* (London: Allen and Unwin, 1986), pp. 20, 128–29.

percent by 1938 (see Figure 2.2).[8] The trend toward higher protection has continued in postwar Japan and has reached heights of agricultural protection unmatched—except by Korea and Switzerland—by other industrial market economies (Anderson and Hayami, 1986; USDA, 1988; Tyers and Anderson, forthcoming). Within the postwar era, independent Korea and Taiwan have compressed the same policy revolution, switching from policies that depressed prices for domestic farmers in the 1950s to heavy protection in the 1980s (Anderson and Hayami, 1986, pp. 17–38; USDA, 1988).

From the time of the Civil War, the United States has conformed to the developmental pattern as well as the usual trade bias. Agriculture in the temperate zone has kept a comparative advantage in the U.S. trade pattern, with cotton and grains serving as traditional exports. American trade policy treated exportable crops badly before the 1930s. Farmers received little or no aid and had to face tariffs on imports, mostly industrial products, ranging from 20 to 45 percent for the half-century from the Civil War to World War I and still around 15 percent in the 1920s (U.S. Tariff Commission, 1934;

---

[8]The slight drop in rice protection between 1918 and 1927 was caused by the failure of Japan to raise the specific rice duty to match domestic price inflation. Similarly, the drops in ad valorem protection during the two OPEC oil shocks were produced by increases in the price denominator rather than drops in nominal duties.

Benedict, 1953; U.S. Census Bureau, 1976, Series U211). Significant income support was not provided until the New Deal. Domestic supports today are generous by previous American standards, though they are lower than those in Europe and Japan (Petit, 1985, pp. 38–56; Anderson and Hayami, 1986; USDA, 1988). In the United States, the drift from taxation toward subsidization since the Civil War took the form of little trade protection throughout, declining tariffs on importable industrial goods, and rising direct subsidy payments after 1933.

Like the United States, Britain, starting in the 1860s, fits the developmental pattern of agricultural policy. From then until 1932, the country had virtual free trade and low tax rates that were nearly neutral toward agriculture.[9] For the rest of the 1930s, there was little shift toward net protection of agriculture. Imports from the Commonwealth were not discouraged, and there were significant duties on steel, autos, and other industrial goods. Britain entered a period of heavy agricultural protection mainly after World War II. It peaked around 1972, when Britain began to conform to the Common Agricultural Policy of the European Community and imposed higher duties on traditional agricultural imports from the Commonwealth. The inflationary oil shock of 1973–74 lowered ad valorem rates in Britain as elsewhere, but by 1980 the protection rates of the 1972 peak had been regained.

Prussia and unified Germany also followed the development pattern from the 1860s on, as illustrated by the rates of protection for wheat and other foodstuffs in Figures 2.3 and 2.4 and Tables 2.2 and 2.3. Germany became an early international leader in the protection of the grain sector with Bismarck's famous tariff package, the "compact of iron and rye," in 1879. Throughout the 1880s, pressure to protect agriculture, as well as essential German values and national security, continued to build. In 1902, after an interlude of freer trade under Leo von Caprivi, the intense pressure from the Junkers and the Bund der Landwirte resulted in duties on grain matching those on protected industries and a virtual prohibition of foreign livestock and meat.

[9]The most noteworthy return of Britain to agricultural protectionism before 1930 was the enactment of a 165 percent sugar production bounty in 1925. The bounty supplemented the protection given by the longstanding duty on foreign sugar, which was equivalent to 100 percent ad valorem in 1925 (and 87 percent in 1927, the year used for other comparisons below). See U.S. Secretary of Agriculture (1933, pp. 267, 284, 521). Britain also had longstanding duties on imports of goods it did not produce such as tea and tobacco. These receive no attention here, because the import duty cannot be distinguished from a consumption tax in such cases.

Figure 2.3. Nominal rates of protection against wheat imports into France and Germany, 1854–1980

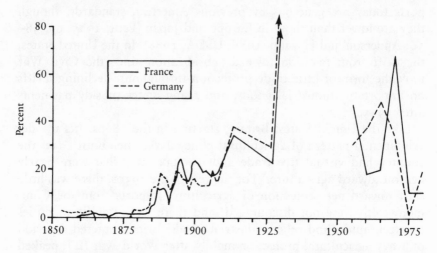

Sources: For 1854–1908, Great Britain, Board of Trade, "British Trade and Industry," in House of Commons, *Sessional Papers,* vol. 102 (1909), p. 908; for 1913, 1927, and 1931, Heinrich Liepmann, *Tariff Levels and the Economic Unity of Europe* (London: Allen and Unwin, 1938), pp. 64, 68; for 1955–80, Kym Anderson and Yujiro Hayami, *The Political Economy of Agricultural Protection: East Asia in International Perspective* (London: Allen and Unwin, 1986), p. 137, subtracting unity from NPC. See also Odd Gulbrandsen and Assar Lindbeck, *The Economics of the Agricultural Sector* (Stockholm: Almqvist and Wiksell, 1973), pp. 33–38, 182, 257.

Note: Annual data used for 1854–1908, 1913, 1927, 1955, 1960, 1965, 1970, 1975, and 1980, with 1931 off scale (180 for France, 212 for Germany).

Protection for industrialists reduced the effective rate of protection for agriculture in imperial Germany. In 1913, for example, tariffs of about 21 percent for foodstuffs were partially offset by tariffs averaging 10 to 15 percent for industrial goods (Table 2.3). The protectionist trend resumed under the Weimar republic in the 1920s and was accelerated by the Nazi march to autarky (Gerschenkron, 1943; Kindleberger, 1951; Gourevitch, 1977; Tracy, 1982, pp. 87–110, 193–216). Support for West German farmers following World War II was less extreme than under the Third Reich, but it remained stronger than the support from the prewar empire or the Weimar republic, peaking around 1965 and declining moderately to 1980 before rising again (see Figure 2.4).

France also tended to follow the developmental pattern after 1860, though the movement from taxing to subsidizing was unsteady. Starting from virtual free trade in the 1860s, agricultural protection rose

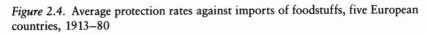

Figure 2.4. Average protection rates against imports of foodstuffs, five European countries, 1913–80

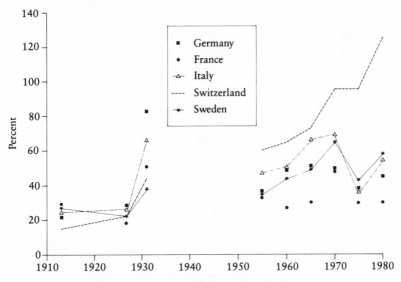

Sources: For 1913–31, Heinrich Liepmann, *Tariff Levels and the Economic Unity of Europe* (London: Allen and Unwin, 1938); for 1955–80, Kym Anderson and Yujiro Hayami, *The Political Economy of Agricultural Protection: East Asia in International Perspective* (London: Allen and Unwin, 1986), p. 26.

Note: For 1913–31, the estimates are Liepmann's "potential tariffs," simple averages of rates of duty on all available products. He preferred simple averages on the grounds that trade-weighted averages would be even worse, giving no weight to prohibitive duties, large weight to zero-percent duties, and so on.

The bundle of foodstuffs for 1955–80 is not the same as those in Liepmann's estimates for 1913–31.

As a check against biases resulting from changing aggregation, I examined the separate rates of protection on wheat and sugar for the same countries and dates. Aside from being noisier than the all-foodstuffs aggregates, the wheat and sugar series yield the same general patterns.

after 1881, but these movements were partly offset by contemporaneous increases in industrial tariffs. Even the Méline tariff of 1892, sometimes cited as a triumph for agriculture, was part of a larger package of duties that actually gave almost as much nominal protection to industry (Golub, 1944; Augé-Laribé, 1950; Tracy, 1982, pp. 59–86, 173–92). This tension between moderate agricultural and industrial tariffs continued as late as 1927 (see Figures 2.3 and 2.4, Table 2.3). In the 1930s, France shared in the global retreat to autarky and self-sufficiency. Postwar French support for agriculture resembles that of West Germany: less autarkic than in the 1930s but

Table 2.2. European tariff rates for wheat and raw sugar for 1913, 1927, and 1931

| Country | GNP per capita in 1960 U.S. dollars | | Tariff rate (percent) | | | | | |
| | | | Wheat | | | Raw sugar | | |
	1913	1929	1913	1927	1931	1913	1927	1931
Germany	731	795	38.0	29.0	212.0	91.5	31.6	280.0
France	695	983	34.5	23.0	180.0	125.0	46.0–54.0	200–240
Italy	441	517	41.5	27.0	144.0	346–390	27–40	195–290
Belgium	894	1,099	0.0	0.0	0.0	79.0	29.0	100.0
Switzerland	963	1,265	1.7	2.0	5.7	20.0	5.1	140.0
Austria[a]	681	720	36.0	1.1	72.0	21.5	43.0	218.0
Czechoslovakia[a]	524	586	36.8	16.5	89.5	21.5	133.0	366.0
Sweden	680	897	8.6	10.0	26.0	43.5–60	26.4–37	68–97
Finland	520	590	0.0	0.0	0.0	197.0	117.0	350.0
Poland[b]	n.a.	n.a.	0.0	0.0	100.0	290.0	74.0	370.0
Romania	n.a.	n.a.	n.a.	n.a.	n.a.	100.0	67.0	195.0
Hungary[a]	372	424	27.6	9.0	69.0	79.0	51.0	140.0
Yugoslavia	284	341	27.6	9.0	69.0	79.0	51.0	140.0
Bulgaria	263	306	2.8	5.4	42.7	100.0	127.0	350.0
Spain	367	455	29.2	19.6	71.0	312.0	153–216	420.0

Sources: Data for tariff rates, Heinrich Liepmann, Tariff Levels and the Economic Unity of Europe (London: Allen and Unwin, 1938); for GNP per capita estimates, Paul Bairoch, "Europe's Gross National Product, 1800–1975," Journal of European Economic History 5 (Fall 1976): 307.
[a]For 1913, the rates for Austria, Hungary, and Czechoslovakia are those for the Austro-Hungarian Empire.
[b]For 1913, the rates for Poland are those of the Russian Empire.

Table 2.3. European tariff rates for foodstuffs and manufactured industrial goods for 1913, 1927, and 1931

| Country | GNP per capita in 1960 U.S. dollars | | Tariff rate (percent) | | | | | | | |
| | | | All foodstuffs | | | Manufactured industrial goods | | | |
	1913	1929	1913	1927	1931	1913	1927	1931
Germany	731	795	21.8	27.4	82.5	10.0	19.0	18.3
France	695	983	29.2	19.1	53.0	16.3	25.8	29.0
Italy	441	517	22.0	24.5	66.0	14.6	28.3	41.8
Belgium	894	1,099	25.5	11.8	23.7	9.5	11.6	13.0
Switzerland	963	1,265	14.7	21.5	42.2	9.3	17.6	22.0
Austria[a]	681	720	29.1	16.5	59.5	19.3	21.0	27.7
Czechoslovakia[a]	524	586	29.1	36.3	84.0	19.3	35.8	36.5
Sweden	680	897	24.2	21.5	39.0	24.5	20.8	23.5
Finland	520	590	49.0	57.5	102.0	37.6	17.8	22.7
Poland[b]	n.a.	n.a.	69.4	72.0	110.0	85.0	55.6	52.0
Romania	n.a.	n.a.	34.7	45.6	87.5	25.5	48.5	55.0
Hungary[a]	372	424	29.1	31.5	60.0	19.3	31.8	42.6
Yugoslavia	284	341	31.6	43.7	75.0	18.0	28.0	32.8
Bulgaria	263	306	24.7	79.0	133.0	19.5	75.0	90.0
Spain	367	455	41.5	45.2	80.5	42.5	62.7	75.5

Sources: See Table 2.2.
[a]For 1913, the rates for Austria, Hungary, and Czechoslovakia are those for the Austro-Hungarian Empire.
[b]For 1913, the rates for Poland are those of the Russian Empire.

more protective of agriculture than in the 1920s or earlier, with a drop in protection in the 1970s and a rise in the 1980s (Delorme and André, 1983, pp. 303–29).

Similar trends were followed by other European governments for which there are data back to 1913. Postwar farmers received more protection in Italy, Sweden, Austria, and especially Switzerland than did their prewar and interwar predecessors (see Figure 2.4) (Liepmann, 1938; Gulbrandson and Lindbeck, 1973, p. 38).

Developing Countries. Agricultural protection grew in each of the now industrialized nations since the 1860s. Was there a similar trend in the long independent countries of the Third World? Did they tax agriculture even more heavily in the past? Historical data are scarce on this issue. For the postwar era, only tentative judgments can be made about trends in the relative taxation of agriculture in developing countries. Over periods of ten years or longer starting from the 1960s, Brazil, Colombia, Korea, Mexico, the Philippines, and Taiwan have shifted to lower taxation or positive subsidization of their farm sectors. From the 1970s to 1981–83, a group of thirteen African nations switched to net subsidization of cereals, with no net trend in the taxation of their export crops (World Bank, 1986, pp. 62, 68). Preliminary results from eighteen developing countries, however, show no change in protection rates from 1975–79 to 1980–84 (Schiff and Valdés, forthcoming). It is still premature to say there was a postwar trend toward lighter taxation of agriculture in developing countries.

It is unlikely that governments of developing countries have been drifting toward lighter taxation over a century or longer.[10] Analysis of the historical data suggests that the anti-trade pattern was always strong and that the drift toward protecting agriculture was weak or nonexistent until the nations reached the threshold of industrial country status.[11] Before World War II, developing countries were

[10]There are limits to how severe the net taxation of agriculture could have been in the past. It is apparent in developing countries that even before the limits to taxpayers' endurance come the limits to tax-collecting competence. Underdevelopment means, among other things, that the government cannot raise more than a tiny share of national product in tax revenues. Surely the history waiting to be written in this area will show how the state gained taxation powers, which happened to be aimed primarily at agriculture.

[11]For cross-sectional studies, see Fetter and Chalmers (1924); Great Britain, Committee on Industry and Trade (1927); and Wright (1935). For national studies, see Brigden et al. (1929), Díaz-Alejandro (1970), and Ingram (1971). For long time-series studies, see, for example, Mitchell (1981) and Wilson (1983).

generally exporters of agricultural products. They followed the anti-trade pattern faithfully, protecting nascent industries with tariff rates of 15 to 35 percent, while denying such support to farmers or mineral exporters. The anti-trade pattern among developing countries remained fixed and trendless from the earliest nineteenth-century data to the present (except in countries such as Egypt, Meiji Japan, or Thailand, where pressure from the great powers delayed the rise of protectionism). There was a countercyclical tendency to stiffen the anti-trade policies during the depression of the 1930s and to weaken them during the inflation of the 1970s, but there was no trend toward protecting agriculture.

Early Patterns in Europe. Indirect clues to the origins of the willingness of developing countries to tax agriculture can come from the history of an earlier, less-developed Europe. In 1860, the only widely used policy instruments were trade policy and direct taxes such as land taxes. The main policy stance of most European nations in early modern times was one of "provisioning," in which the state was prepared to compel affordable deliveries of necessities, especially of food and especially to the cities (Heckscher, 1934, 2:80–111; Kaplan, 1976; Outhwaite, 1981; Fogel, 1989, pp. 28–35). The instruments chosen involved little budgetary outlay, often resulting in weak enforcement. During grain shortages, the throne would issue edicts demanding that grain be delivered at prices at or under a decreed maximum. Sometimes local officials were compelled (at little or no royal expense) to inventory private grain reserves for possible state seizure. Exports were often banned. The clear intent was to make the food supply system pay the price of food insurance for its customers. In this sense, it was a policy that taxes agriculture (and merchants), albeit less harshly than do many of the agricultural policies in developing countries today.

By far the most important exception of provisioning, however, was England under the Corn Laws (1660–1846). As Bates (1988) has rightly argued, England in this era was so exceptional that her dominance in the historical view of the English-speaking world has distorted our understanding of long-run trends in policies affecting agriculture.

Tudor and early Stuart monarchs strove to follow the provisioning policy, especially in times of crisis. During the period 1600–1640, their efforts seem to have succeeded in the sense of noticeably reducing the deviations in grain prices from trend (Fogel, 1989, p. 34).

Table 2.4. Differences between the English and Continental silver prices for wheat, by decades, 1621–1910

Decade[a]	The English price compared with the Continental price in terms of the Continental price[b]			
	Belgium	France	Amsterdam	Northern Italy
1620s	1.6	−13.9		−29.2
1630s	−2.7	−11.2		−0.4
1640s	1.6	5.0		15.7
1650s	0.4	−12.1		42.9
1660s[c]	14.5	−1.0		−59.4
1670s	10.8	33.3		38.9
1680s	9.7	16.6		46.1
1690s	7.6	21.6		94.7
1700s	−0.4	6.9		34.9
1710s	35.0	36.3		64.9
1720s	55.3	37.4		95.1
1730s	33.1	16.9		29.6
1740s	12.3	23.3		8.0
1750s	57.8	39.0		41.6
1760s[d]	57.8	46.2		55.6
1770s	55.4	30.1	26	29.1
1780s	43.7	27.6	25	26.8
1790s–1810s: Wars, inconvertible sterling, and suspension of Corn Laws				
1820s[e]		37.6	44	71.2
1830s		28.0	29	51.4
1840s[f]		18.5	24[g]	23.5
1850s		9.6		10.0
1860s		0.5		9.2
1870s		−4.9		
1880s		−17.7		
1890s		−27.6		
1900s		−26.2		

Sources: Prices for Belgium, France, Northern Italy, and Hamburg, Wilhelm Abel, *Crises agraires en Europe (XIIIᵉ–XXᵉ siècle)* [Agrarian crises in Europe: The thirteenth to the twentieth century] (Paris: Flemmarion, 1973); prices for England, author's own calculation based on those through 1770 at Eton College. Those from 1771 onward are the England-Wales average from Brian Mitchell, with the collaboration of P. Deane, *Abstract of British Historical Statistics* (Cambridge: Cambridge University Press, 1962), pp. 486–89; prices for Amsterdam are those for Polish wheat from N. W. Posthumus, *Inquiry into the History of Prices in Holland* (Leiden: E. J. Brill, 1946), adjusted for weights and measures and for exchange rates.

Note: For Amsterdam prices, the unit of measure, a "last," was apparently 10.89 quarters or 87.125 bushels of grain weight, as explained in Postlethwaite's *Dictionary of Commerce* (1774), kindly supplied by Larry Neal. Dutch currency values had to be converted into sterling at different rates before and after the start of 1827. For the time after 1827,

The regulatory machinery was dismantled under Cromwell. After 1660, the Restoration not only failed to reassemble the provisioning machinery but took a very different tack. The provisioning policy had been aimed at eliminating high grain prices during shortage, with little intended effect on prices during normal years. The new Corn Laws, by contrast, were aimed at raising grain prices in normal years, turning passive and permissive only in the years of greatest shortage. To this end, the government paid export bounties in non-crisis years over the century when England remained a net grain exporter as often as not (1660–1765). Import duties were also decreed. The import duties were frequently adjusted, especially after their relevance was enhanced by England's permanent shift after 1765 to being a food importer. What evolved was a system of sliding scales, in which the lower the latest domestic grain prices, the higher the duty on imports.

The price-raising effect of the Corn Laws increased over the century from 1660 to 1765 when England was a net exporter (see the comparison of English and Continental wheat prices in Table 2.4). After 1765, being a net grain importer allowed England to protect her agricultural interest and simultaneously raise revenue by taxing imports. As Table 2.4 and Figure 2.5 show in complementary ways price-propping in England was effective in the 1765–1846 era, and the import duties seem to be the reason. The policy remained flexible, however. In an extreme crisis in food supplies, starvation and rebellion

Table 2.4. (continued)

Posthumus (1946) gives consistent rates for guilders (fl.) per pound, always near 12. Earlier quotes were in banco Flemish schillings per pound. Sir Isaac Newton is quoted by Postlethwaite (per Larry Neal) to the effect that 20 banco Flemish schillings = 6 guilders. Wheat prices before 1827 were thus converted into sterling by multiplying 20/6 schellingen per guilder by the wheat price in guilders per last, divided by the schellingen-to-pound exchange rate, and then converted from pound per last to pound per quarter. In comparing English prices to those of Amsterdam, the figures refer to calendar years until 1815. Thereafter they refer to years from "1st July to 11th March [sic]."

[a]For Belgium, France, and Northern Italy, decades are defined so that, for example, 1820s = 1821–30, whereas for Amsterdam and Hamburg, 1820s = 1820–29.
[b]Each figure in the table is calculated as 100 × (England−other)/(other).
[c]English Corn Laws begin.
[d]England becomes an importer.
[e]For the 1820s, the difference between the Hamburg price and the English price is 110.4.
[f]Corn Laws end.
[g]1840–45 only.

Figure 2.5. British Corn Law duties and the England-Amsterdam price differential on wheat, 1766–1860

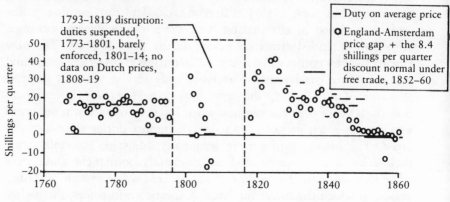

Source: See Table 2.4.
Note: The duty levied is in shillings per quarter, which is equivalent to 8 bushels. The rates of import duty were inferred from the discussion in Donald G. Barnes, *A History of the English Corn Laws* (New York: Augustus Kelley, 1930; rpt. 1961), pp. 250–52; and C. R. Fay, *The Corn Laws and Social England* (Cambridge: Cambridge University Press, 1932), chaps. 2 and 3. They can only be approximations, given the odd timing of changes in the official prices and their interaction with the nonlinearities of the sliding scales. The author's procedure was to find the duty corresponding to the annual average English price.

had to be avoided. In the French War era (1793–1815) and in other years of extreme hardship, the Corn Laws were either suspended or implemented with cautious partiality (Barnes, 1930; Fay, 1932).

What Do the Historical Patterns Show? History thus sharpens our perspective on comparative patterns of agricultural policy in different ways. In the preceding discussion, the developmental pattern and the anti-trade bias are extended back to the 1860s.[12] Yet they show a different profile in history than in the modern global view. Figure 2.6 underscores the difference by drawing the two patterns schematically. Figure 2.6a sketches the general international tendencies today in terms of an effective rate of protection, one imagined to capture all general-equilibrium effects. The switch from taxing to subsidizing is common to agricultural importables and exportables, though the former are always less taxed or more protected.

[12]For want of data, the commodity pattern is harder to test before World War II than are the anti-trade and developmental patterns. The only abundant data comparison is that between sugar and wheat protection for 1913–31. Table 2.2 makes it clear that sugar was more heavily protected than wheat throughout Europe in this era, whereas NPCs for sugar only slightly exceeded NPCs for wheat among industrial countries in the postwar era.

Figure 2.6. Summary of global and historical patterns of agricultural protection

(a) *The global cross section, 1980*

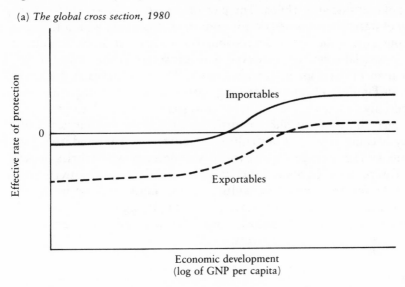

(b) *Historical paths since the 1860s*

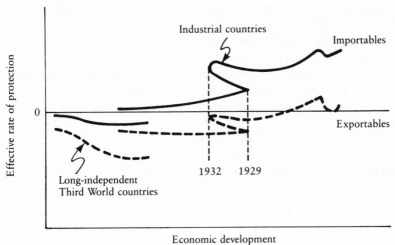

Figure 2.6b shows the somewhat different story from the history of policies since the 1860s. The pair of curves on the right traces the rise of agricultural protection in industrial countries, with a lurching detour during the Great Depression. The early taxation of agriculture by industrial countries, however, was less severe in the 1860s or earlier than in developing countries today. The two curves at the lower left in Figure 2.6b show that long-independent developing countries, especially in Latin America, were consistent in their relative taxation of agriculture. They did not drift toward protecting farmers until they became either net agricultural importers or truly industrialized nations. The average experience of still-developing countries shown in Figure 2.6b combines this static Latin American pattern with a drift toward taxing agriculture a bit more heavily in a few agricultural-exporting countries (for example, Egypt and Thailand) whose ability to favor industry was checked until this century by pressure from the great powers.

Explaining the Patterns of Agricultural Protection

A difficult task has now been set. We must now explain dynamic paths of agricultural policy, not just static cross-sectional puzzles. The historical record demonstrates a richness and variance in these paths that prohibit the simple extension of "development" variables as explanatory factors in the model. How, then, do we explain the similarities and differences between the patterns of history and those of today? How do we deal with exceptions like Corn Laws England? To answer these questions about the general patterns and their exceptions, we turn first to some familiar explanations and then to some new ones.

Popular Explanations: Rhetoric and Reality

A typical reaction to the developmental pattern of agricultural policy is that the basic explanation is easy and well-known. There are in fact several explanations that are well known and commonly invoked—and inappropriate. They have a common shortcoming: they are all theories simply of the urge to protect. They cannot account for the equal or greater prevalence of discriminatory taxation of agricultural producers or for the switch from taxes to protection.

'Food Security. The most common rationale for supporting farm incomes and imposing barriers to food imports is the need to assure a safe food supply for future crises. The rhetoric of food security was prominent in the building of the strongest farm supports, especially in Japan and Switzerland, two nations scarred by memories of shortages during World War II.

Among the high-income countries, concerns for food security cannot explain the degree and pattern of agricultural protection so readily observed. Rice policy in Japan is a case in point. Since 1968 the government of Japan has been disposing of surplus rice beyond the amount the nation wants to stockpile for emergency reserves. Japan has even become a net exporter of rice since 1977. Japanese rice policy is not the only case of a glaring departure from goals of food security. Canada, the United States, and the EC also, like Japan, subsidize acreage reductions and exports of food crops. EC dairy products, like Japanese rice, were converted from importables to exportables by generous subsidies to producers. The types and levels of farm subsidies among the high-income countries reveal that support of farm incomes, not food security, is the main motive.

In developing countries, the set of governments with the ostensible goal of food security far exceeds the number that protect the food sector. Out of twenty-five combinations of nations and food products for which nominal protection coefficients were measured for the period around 1980, only four were cases of genuine protection (NPC above 1.20) (Binswanger and Scandizzo, 1983). Seven others were nearly neutral (NPC between 0.80 and 1.20), and fourteen taxed food producers severely (NPC < 0.80).

Agricultural Price Instability. The agricultural products sector is afflicted with particularly unstable prices and perhaps also with particularly unstable producer surplus.[13] Policies to raise farm incomes are often defended as insurance against the problem of price instability.

Yet the policy and the problem are a mismatch in three ways. First, they are a conceptual mismatch, because instability calls for stabilization policies, not a perennial income redistribution. Second, they are a historical mismatch. To match the problem and the demand for support policies over time, one should show rising instability of

---

[13]For a useful survey of instability, as well as trends, in the terms of trade for primary products, 1900–1982, see Scandizzo and Diakosawas (1987).

producer surplus (or at least domestic prices) in less regulated settings, followed by lower instability after support policies were enacted. This is not the trend pattern (Scandizzo and Diakosawas, 1987, pp. 58–103, 164–65). Only in the Great Depression of the 1930s were instability and public demand for support policies linked.

Finally, the problem of instability lacks a strong raw correlation with farm-support policies when we look across the spectrum of commodities. The most protected commodities are not the ones with the greatest price deviations from trend since 1900. Among those with the worst instability in the terms of trade, only sugar, highly protected and highly unstable in price, favors the instability explanation of support policies. Even in the case of sugar, it is not clear that reverse causation, from the policy to the instability, can be ruled out, especially since high levels of sugar protection started well before the 1930s and became more generous as price instability rose over time. Furthermore, among the primary products with the worst price instability were three tropicals that are now heavily taxed: rubber, cocoa, and coffee (Scandizzo and Diakosawas, 1987, p. 71).

The price instability argument needs to be more carefully specified and tested. It will be repackaged as part of a "rising income sensitivity to price" theme in the next section.

Agricultural Protection as Poor Relief. A common intuition, though one seldom made explicit by scholars, is that agriculture gets special policy protection because farmers would have income far below the average if they were not supported by government. It is plausible to view the rise of generous farm supports as part of the broader twentieth-century mandate to fight poverty. The idea is not refuted by the fact that median income is as high for U.S. farm households as for other U.S. households and for Japanese farm earners, almost as high as for all Japanese earners (Hayami, 1988, pp. 20–21, 92–93). It could still be true that the average farm earner would be poorer than the average earner if the generous postwar supports were removed. Nor can the political perception of this point be dismissed, even though the actual benefits of farm supports eventually accrue to the wealthier owners of farmland, not to the average or small farmer.

The main stumbling block in the path of the poor-relief interpretation of the developmental pattern in agricultural policy is that higher-income settings are not the ones in which farmers have their lowest earning potential aside from government support. The best crude indicator of the part of farm earning power that is less directly tied to government subsidy is gross product per employed person in agriculture, compared with gross product per employed person outside of agriculture.[14] This indicator did not fall over time or development during the shift from agricultural taxation to agricultural protection. In the international cross section of today, this relative labor productivity of agriculture is slightly higher in the most industrialized, and farm-protecting, countries, failing to show their greater need to offset agricultural poverty. The picture is similar over the long sweep of history: the relative labor productivity of agriculture has not declined in the United Kingdom since 1856, or in the United States since 1900, or in Japan since the Meiji reign.[15]

Despite the clear link between farm poverty and the demand for support policies in the depths of the Great Depression, there is reason to doubt that trends in the threat of farm poverty explain trends in policy. Indeed there is an equally plausible counterargument suggesting that farm poverty should weaken political support for farmers. The greater their social distance, and particularly their income distance, from the rest of the population, the less sympathy there may be for subsidizing farmers on self-insurance grounds ("I could end up in their shoes"), as suggested in the pressure-group framework. This relative-poverty argument is tested statistically in the next section.

Nostalgia. Modern nonfarm voters respect the farm life enough to pay for its support through government (though not enough to live on a farm). Although the nature of the respect varies, there is no denying that distant glow. Yet it too fails to help explain why government support for agriculture rose. Reverence for the agricultural life has such a long history in terms of rhetoric that the sentiment must have been widespread centuries before the onset of farm protection and subsidies. One could argue that it lacked political power until the nostalgic urban classes reached critical size. The size argument, however, clashes with the consensus that smaller group size is a political advantage. Also, if the spread of nostalgic pro-farm sentiment

[14]The relative labor productivity of agriculture (r_a) has been derived from data on the share of gross national product originating in agriculture (y_a) and the share of agriculture in total employment (n_a) using the formula $r_a = (y_a/n_a) \cdot (1 - y_a)/(1 - n_a)$. The data on y_a cannot be pure of all direct influences of government policy. Being shares at current prices, they reflect government manipulation of the terms of trade. They do succeed in omitting subsidy payments, however.

[15]For the United Kingdom, see Feinstein (1972). For the United States, see U.S. Census Bureau (1976). For Japan, see Hayami (1988), pp. 20–21, 92–93.

was indeed correlated with economic development and with protective policies, the best bet is that it is an endogenous by-product of the development process and not an independent explanatory force.

A Simple Model of Pressure-Group Interactions

To bring order to the economic influences on agricultural policy, let us consider a modest and frugal theoretical framework. It should be modest because the political process is complex. It should be frugal to concentrate on those few basic forces that transcend that complexity. Yet a framework there should be, to keep our imaginations logical.

Four variables provide the core of the model.[16] The driving force is the potential to use government interventions to redistribute economic benefits away from groups determined by market processes and toward groups selected by government policy. In the model, the amount redistributed is G, and the size of the subsidized and taxed groups are N_S and N_T, respectively. The welfare losses resulting from economic distortions are also shared by these two groups and are equal to D_S and D_T, respectively. The potential for voters to express concern for the welfare of other groups is reflected in two "caring" coefficients, b and c, that measure the extent of sympathy (at the margin) for the subsidized group and taxed group, respectively.

The predicted impact of each set of variables on the amount of redistribution (G) is summarized in Table 2.5. Although finding suitable variables to measure accurately the influence of each component of the model is difficult, Table 2.5 does indicate that clear-cut predictions about the impact of several observable variables can be made. The logic of each prediction, the difficulty of measuring its effect, and the nature of counterhypotheses are discussed in the following sections.

Sector Size and the Developmental Pattern. The front-running hypothesis for explaining how farmers evolved from beasts of burden to favored pets is the "small is powerful" argument. Its logic is compelling. Part of the logic is that appeasing a smaller group is cheaper.

---

[16]The model presented in this section and the Appendix is a condensation of that in Kristov, Lindert, and McClelland (1989). Other historical applications of the same model can be found in Lindert (1989). It draws on earlier economic models of pressure group competition, particularly Becker (1983) and (1985). For a critical comparison of these and other models of the political process with applications to agricultural policy, see Alston and Carter (1989).

Table 2.5. Predicted influences on redistribution through government, according to the simple pressure-group model

Parameter	Effect of raising parameter on redistribution[a]	Observable measures of parameter
Marginal deadweight loss from extra redistribution, borne by		Administrative costs,
the subsidized group	Negative	elasticities-based side
the taxed group	Negative	(deadweight) cost formulas
Individuals' marginal caring about		Attribute "distances" of
subsidized group	Positive	individuals with voice from
taxed group	Negative	the affected group
Size of subsidized group	Negative	The group size itself
Size of group to be taxed	Positive	The group size itself

Some applications:

1. The rise of government is limited by the exhaustion of positive-sum programs and the rise of deadweight losses.

2. For given sympathies, a proposal aimed at smaller affected groups evokes more intense political support. The faster drop of group size for agriculture than for other sectors raises its lobbying power faster than for other sectors.

3. When political voice is concentrated in a small minority, that minority behaves like a price-discriminating monopolist toward the relatively voiceless masses. Groups more likely to exit when taxed (by rebelling, emigrating, evading taxes, or dying) are taxed less (e.g., the text's discussion of the relative rise of rural voice and of Corn Law Britain).

4. Lower price elasticities mean lower national deadweight losses from extra redistribution. The fact that price elasticities are lower for agriculture than for other sectors of similar size magnifies the amounts of redistribution either from or to agriculture. In early settings it magnified the taxation of agriculture, and later it magnified the protection of agriculture.

5. The same elasticities argument predicts part of the protection ranking of agricultural commodities in most industrialized countries today: dairy > beef > sugar ≥ grains > (pork, chicken, and eggs). Yet regression tests do not support this prediction.

[a]Other things being equal.

When farmers were half the voting population, a $100 pure transfer to each farmer would cost the average nonfarmer $100, and there would be strong resistance. When farmers had become only 5 percent of the population, giving $100 to each farmer cost the average nonfarmer only $5.26 (= 5/95). Mobilizing a smaller group is also cheaper. As long as there are fixed costs to becoming politically active (and the distributions of individual stakes are similarly shaped in the two opposing camps), the larger group will suffer more free riding and inactivity (Olson, 1965, 1985). Such simple arithmetic cannot be the whole story, but it must have played at least some role in the drift toward farm subsidies in every country.

Rising Income Sensitivity to Prices. The voices of farmers may also have been raised by a long-term upward trend in their sensitivity to shifts in prices, shifts that government could control. Here lies a potential explanation for the developmental drift from taxation to protection.

Price sensitivity has been raised by a simple basic fact of economic development: farm products decline as a share of farmers' own consumption. In early settings, the farm sector consumed a large share, say half, of its own product, including housing (Brady, 1972, p. 80; Gregory King [1695] in Laslett, 1973, pp. 65, 210). A drop of 10 percent in the price of all farm products would lower the real income of the farm sector by only 5 percent. Today, by contrast, the farm sector sells 90 percent or more of its gross income in exchange for nonfarm products, so that a drop of 10 percent in the price of all farm products would lower the real income of the sector by 9 percent. The primary source of the rise in the "openness" of the farm economy is, of course, Engel's Law that the share spent on nonfood items rises with the level of income. Another source is the decline in transport and communications costs, which integrates individual farms into the market economy.

The rise in market exposure can be quantified at the level of individual farms for the twentieth century, with only rough suggestions of an earlier rise. In the United States, the marketed share of gross farm product had risen to 78 percent by 1910 (see Figure 2.7). An earlier increase in exposure to the vagaries of the market is believed to have been a key to the puzzle of U.S. farm protest in the midst of favorable farm-income trends in the late nineteenth century (Mayhew, 1972; McGuire, 1981). In the twentieth century, there has been a further rise in the share marketed by individual farm households, to around 90 percent of household gross income (see Figure 2.7).[17] It also seems likely that value added as a share of gross output has fallen faster in agriculture than in industry, raising income sensitivity to price changes.

[17]Additional estimates of the shares of U.S. farm expenditures devoted to the same farm's product can be found in Funk (1914) and Schultz (1953, p. 240). The suggestion here that the share was about half in early stages of development was based on fragmentary U.S. and English data. Lower marketed shares could be found in other settings, of course. See, for example, the very low trade shares for farm households in Tokugawa-era Japan in the 1840s estimated by Smith (1969). The distinction between farm incomes gross and net of purchased farm inputs is ignored here. Net incomes are more appropriate. But the available data relate to gross income.

Figure 2.7. The open farm economy: The share of U.S. gross farm income derived from off-farm sales of product, 1910–84

Source: Calculated from Gary Lucier, Agnes Chesley, and Mary Ahearn, *Farm Income Data: A Historical Perspective*, Statistical Bulletin No. 740 (Washington, D.C.: U.S. Department of Agriculture, Economic Research Service, 1986), p. 14.

Another possible factor in rising sensitivity is the magnification effect of commodity prices on farm income, which seem to rise with economic development, whereas the effects of the same commodity prices on the nonfarm sector become smaller. Anderson (1987) found this effect to be very strong in a quantified general-equilibrium model with three sectors and four factors, only one of which was mobile between sectors. The more advanced the economy, he argued, the greater the percentage impact of the terms of trade on farmland returns and the smaller its percentage impact on nonfarm factor incomes.[18]

The Role of Government Demand for Revenue. The anti-trade bias, and possibly the developmental pattern, of agricultural policy must in large measure stem from the fiscal demands of the state. The relentless search by the state for sources of revenue helps explain, in at least three ways, the historical and global patterns of agricultural policy.

[18]Anderson's result is not tested and not confirmed by other plausible models, however. It remains a tantalizing possibility awaiting a test. In a two-sector model with two mobile factors and immobile farmland, the effect of sector size on the income sensitivity of the farm sector to prices is ambiguous.

In any country, the easiest way to explain the anti-trade bias is to acknowledge that the state is a peculiarly strong-voiced, special interest lobby. There is always at least some pressure to raise revenue by taxing imports, exports, or both. Although the degree of pressure to tax trade can vary with the strength of the protected and taxed groups and with the strength of the national security argument, the bedrock explanation for the tendency to favor producers of importables over producers of exportables is the interest of the state in revenue. Indeed, this point is general and not specific to agriculture.

A second explanation can be traced to an "infant government" effect. In the least-developed settings, a nascent state apparatus must raise revenue where it can. It will concentrate on those tax bases that are large, easily monitored, and politically voiceless. In such settings, agriculture qualifies in all respects. It generates a large share of taxable product. Its products are visible, especially in the case of exports, which (like extractive mineral exports) need only be monitored and taxed in a few key ports. And in most cases the agricultural population is poorly mobilized for lobbying. The willingness to impose heavy burdens on agriculture may have been partly a by-product of early state-building, first in early modern Europe and as late as the 1970s in developing countries.

Agriculture versus Other Sectors. The switch to protection is more pronounced for agriculture than for other declining sectors. Other primary sectors (such as mining, forests, or fisheries) were not taxed so heavily when national income per capita was low, and they are not heavily subsidized in industrialized countries. Part of the explanation is easy. Whatever validity the small-is-powerful argument may have, it should have had more effect in agriculture than in other sectors. Agriculture shrank faster as a share of the electorate, the labor force, and GNP than other sectors, thanks largely to Engel's Law. Even if policies were equally responsive to changes in the size of any sector, they would have changed more in agriculture, which shrank more.

Once other forces, such as sector shrinkage, tip the political balance from taxing a sector to subsidizing it, the resulting swing in policy will be more pronounced, the lower the relevant long-run price elasticities. Lower price elasticities mean lower deadweight costs from each extra dollar of revenue redistributed from one group to another. Lower deadweight costs per dollar redistributed tip the political scales in favor of extra redistribution. Abstract as the idea may sound, there is a mechanism for its affecting political debate.

Opponents of a redistribution will be more ardent, and more persuasive to others, the greater the perceived waste and distortion caused by the redistribution. Correspondingly, those in favor of the redistribution will fight less ardently the greater their own participation in the net national deadweight loss, which in turn would be raised by higher elasticities.

Both the early taxation and the later protection of agriculture may have been reinforced by having lower price elasticities and therefore lower deadweight costs to check the forces in favor of (first) taxation and (later) subsidies. The premise that price elasticities are lower for agricultural products seems correct, even though econometric attempts to estimate elasticities are notoriously shaky, often leading to underestimates (Schultz, 1953, pp. 186–94; Askari and Cummings, 1976; Timmer, Falcon, and Pearson, 1983, pp. 104–9; Carter and Gardiner, 1988). On the demand side, the own-price elasticities of demand for most agricultural products (especially foods) are lower than for other sectors of the same size, basically because income elasticities are lower for these products than for nonagricultural products taking similar shares of consumer budgets. On the supply side, low elasticities again seem more dominant in agriculture, perhaps because of the greater reliance on land, a relatively inelastic input. Here may lie a key argument in explaining why the agricultural policy pendulum swings further than that of such other sectors as mining, forestry, fisheries, or textiles.

The Exceptionalism of the British Corn Laws. The greatest exception in history to the rule that the now industrialized countries drifted from taxing to subsidizing agriculture does fit the present interpretive framework, once we note the key peculiarity of that exceptional case. England stood out among early modern nations in its desire to make grain expensive at home from the 1660s to the 1840s, when it became the nation most willing to abandon its agricultural interest in favor of free trade. After 1660, and especially after 1688, England stood out among countries of western Europe as a nation dominated by the landed aristocracy. Indeed, even in the nineteenth century, agricultural landownership was outstandingly concentrated and outstandingly correlated with overall wealth, by international standards (Spring, 1977, p. 6; Lindert, 1986, 1987). Repeal of the Corn Laws was possible in the 1840s largely because the Reform Act of 1832 and other changes in political representation tipped the scales in favor of industrial interests.

Although this core political explanation of English exceptionalism is exogenous to the pressure-group framework, that framework helps explain when and why the Corn Laws were differentially applied. Over time and space, the Corn Laws were flexibly applied in ways that fit a basic economic model of how a political monopoly permits an elite to exploit the rest of society. Parliamentary struggles over the design of the Corn Laws pitted the obvious landed interest in expensive grain against the dangers of exit by a starving or rebellious pauper host. Export bounties and import duties were suspended in isolated years of shortage, when their enforcement would have raised mortality and unrest the most. They were also suspended or only slightly applied throughout the French War era, when scarcity of food and the threat of revolution were perennial. In such times of maximum danger of exit, poor relief also hit its peak. When and where the need to feed the poor was less, the Corn Laws were stiffened. After the Reform Act of 1832 and related changes in local government shifted power toward urban and industrial interests, the Corn Laws were in retreat.

Statistical Tests of the Pressure-Group Model

The above interpretations were based on data that are qualitative or only spotty in their quantification. A consistent quantitative data set can enable further testing of some of the same interpretations and some others. Until better historical data have been gathered, the best data base for formal testing consists of a global cross section around 1980. For 1980 it is possible to compare more countries and commodities and to try out more hypotheses than was done in the regression analyses by Honma and Hayami (1986). Three sources of data (namely, Binswanger and Scandizzo, 1983; Anderson and Hayami, 1986; and Tyers and Anderson, forthcoming) can be combined to yield net protection coefficients for a set of 247 cases of nations and commodities. Of these, 113 cases are drawn from 13 developed countries and 134 cases from 39 developing countries. Fourteen different agricultural commodities are represented.

For the present, nominal protection coefficients remain as the dependent variable. They cast agricultural policy as a single shadow, the international profile revealed by comparing domestic and border prices. There is no better alternative until more estimates of producer subsidy equivalents are available. In what follows, it will be as though the only relevant policies were tax rates on imports and exports because of the necessity to use (logs of) NPCs as the dependent variable.

The testing strategy was to nest some of the competing hypotheses in a single linear equation and to subject each hypothesis to competition against the other hypotheses and competition against a set of what are called here "merely descriptive" variables. A merely descriptive variable is one whose statistical influence on the NPC simply restates a pattern instead of explaining it. For example, the developmental drift from taxing agriculture to subsidizing it is merely restated by showing that a higher GNP per capita raises the NPC.

The Model and Empirical Results

Table 2.6 separates the merely descriptive variables, below the mid-table line, from the variables more directly tied to causal hypotheses. The results are surveyed from the top down, that is, starting with the more directly causal variables.

Sector Size. The share of the labor force in agriculture is a policy influence featured both in past tests and in pressure-group models. Following Honma and Hayami (1986), a quadratic version is explored in the top two rows of each regression, with varying samples and descriptive-variable controls. The conventional prediction is that the protection-maximizing share of farmers in the labor force (and the electorate) is very low, perhaps around 4 to 6 percent. To get such a result, the regression needs a positive slope to N_a/N and a negative slope to its square. Among the developed nations (regressions 2 and 5), this is the result, though not with much statistical significance to the curvature. The slopes imply that the farm sector gets the most protection when it employs 3 to 4 percent of the employed labor force. So far, the results resemble those of Honma and Hayami (1986), which were based on virtually the same sample.

Adding developing countries to the sample might have repeated this result—by adding countries with higher share employed in agriculture and with less agricultural protection. The other four regressions broadly agree by predicting that developed countries will protect agriculture more than will developing countries. Yet the extra regressions complicate matters for the small-is-powerful hypothesis. In no case, beyond the sample of developed countries studied by Honma and Hayami, are the coefficients of sector size significant

Table 2.6. Regressions that predict nominal protection coefficients ca. 1980

Independent variable[a]	(1) 247 country-commodity cases	(2) 113 country-commodity cases, DCs only	(3) 134 country-commodity cases, LDCs only	(4) 247 country-commodity cases	(5) 113 country-commodity cases, DCs only	(6) 134 country-commodity cases, LDCs only
N_a/N = share of agriculture in labor force	-0.38 (1.06)	14.72 (9.44)	0.14 (1.28)	-2.58** (0.88)	4.95 (8.75)	0.93 (1.17)
(N_a/N) squared	-0.12 (0.96)	-177.50* (76.40)	-0.77 (1.12)	1.62 (0.82)	-69.22 (66.98)	-1.40 (0.95)
Democracy and agricultural decline						
DEMOC dummy	-0.19 (0.13)		-0.42* (0.17)	-0.04 (0.12)		-0.25 (0.16)
Agricultural decline, when DEMOC = 1	2.88* (1.05)	4.63* (1.86)	6.43** (1.89)	2.95** (1.02)	3.69* (1.83)	2.49* (0.98)
Agricultural decline, when DEMOC = 0	-0.14 (0.80)		0.06 (0.85)	-0.37 (0.78)		-1.23 (0.73)
$\ln(Y_a/Y_{non})$ = agriculture's relative income per occupied person	-0.01 (0.08)	0.03 (0.13)	0.03 (0.10)	-0.06 (0.07)	-0.004 (0.13)	-0.05 (0.09)
World price instability, exportable	-0.005 (0.01)	0.02 (0.015)	-0.016 (0.014)	0.007 (0.01)	0.016 (0.016)	-0.020 (0.014)
World price instability, importable	0.03** (0.01)	0.022* (0.01)	0.038* (0.015)	0.03** (0.01)	0.025* (0.011)	0.037* (0.016)
Marginal deadweight gain per dollar of producer gain at 10 percent tax rate						
On an exportable (DEADEX)	0.41 (0.28)	0.29 (0.27)	1.63* (0.73)	0.45 (0.28)	0.23 (0.27)	1.28 (0.73)
On an importable (DEADIM)	-0.10 (0.10)	0.08 (0.10)	-0.52* (0.21)	-0.10 (0.10)	0.06 (0.10)	-0.53 (0.21)
Nation independent only since World War II, exportable good	-0.16 (0.19)		0.08 (0.22)	0.29** (0.13)		0.16 (0.15)
Nation independent only since World War II, importable good	-0.41** (0.14)		0.39* (0.16)	0.50** (0.12)		0.23 (0.15)

Two Gerschenkron-hypotheses variables:	(1)	(2)	(3)	(4)	(5)	(6)
National saving rate	0.004 (0.009)	−0.05* (0.02)	0.02 (0.015)	0.007 (0.008)	0.006 (0.011)	0.008 (0.011)
National saving rate times backwardness (i.e., ln [$8,069/GNP per capita]	−0.005 (0.005)	0.35** (0.12)	0.001 (0.008)	0.003 (0.004)	0.020 (0.014)	0.004 (0.006)
Shift terms						
A DC exportable good	−0.33 (0.23)	−0.25 (0.23)		−0.04[b] (0.11)	−0.23 (0.15)	
A DC importable good	0.40* (0.19)					0.04 (0.18)
An LDC exportable good	−0.35 (0.17)		−0.24 (0.21)			
ln [GNP per capita/$4,000] in 4 cases:						
DC exportables	0.17 (0.32)	6.48** (2.29)	−0.43* (0.21)			
LDC exportables	−0.27 (0.16)					
DC importables	0.16 (0.29)	6.60** (2.30)	−0.16 (0.20)			
LDC importables	0.11 (0.14)					
TA = data from Tyers and Anderson[c]	0.27** (0.08)		0.18 (0.10)	0.24** (0.08)	0.14 (0.10)	0.14 (0.10)
Constant	0.36 (0.29)	−3.79** (1.31)	−0.98* (0.44)	−0.02 (0.23)	−0.18 (0.34)	−0.71 (0.33)
Adjusted R²/standard error of estimate	.48/.35	.34/.30	.48/.37	.45/.36	.29/.31	.46/.38

Sources: Data come from 13 developed countries and 39 developing countries. The nominal protection coefficients are from Kym Anderson and Yujiro Hayami, with associates, *The Political Economy of Agricultural Protection: East Asia in International Perspective* (London: Allen and Unwin, 1986); Hans P. Binswanger and Pasquale L. Scandizzo, "Patterns of Agricultural Protection," World Bank Discussion Paper ARU 15 (Washington, D.C.: World Bank, 1983); and Rodney Tyers and Kym Anderson, *Distortions in World Food Markets* (Cambridge: Cambridge University Press for the Trade Policy Research Centre, forthcoming). Data on independent variables are generally taken from official international sources, especially the World Bank. The world price instability measures are sums of absolute deviations from log trend for 1960–72 from Elio Lancieri, "Instability of Agricultural Exports: World Markets, Developing and Developed Countries," *Banca Nazionale del Lavoro Quarterly Review* 131 (1979): 287–310, and receive support from Pasquale L. Scandizzo and Dimitris Diakosawas, *Instability in the Terms of Trade and Primary Commodities, 1900–1982*, FAO Economic and Social Development Paper 64 (Rome: Food and Agriculture Organization of the United Nations, 1987).

Note: See Appendix for notes on Table 2.6.
[a]Dependent variable = ln(NPC), so that + = protected and − = taxed; standard errors in parentheses.
[b]Both DC and LDC exportable good.
[c]Tyers and Anderson (1986) and (forthcoming).
*Significant at the 95 percent level (two-tail).
**Significant at the 99 percent level.

singly or jointly. There are two warnings here. First, sector size alone does not reliably explain the developmental pattern, and it needs help from other forces, probably forces affecting either the efficiency of the farm lobby or the sympathy it gets from the rest of the population, or both. Second, the extra forces to be added seem to have very different effects in developing countries.[19]

Democracy and Agricultural Decline. Two other intuitions about the rise of agricultural protection work well only when they are combined. One intuition, supported by raw correlations, is that the voices of farmers may be better heard in an electoral democracy. The other is that sympathy for protecting agriculture is kindled by rapid agricultural decline. Rapid decline could raise fears that a way of life is disappearing, or it could give the farm sector disproportionate political representation (for example, by leaving more elected representatives per capita in farm areas).

The regressions recommend combining the two ideas. In cases in which the share of agriculture in the labor force did not decline much between 1965 and 1980, agriculture did not gain any more protection in 1980 under electoral democracies than without democracy. Without democracy, a more rapid decline of agriculture similarly failed to raise protection. Yet within the set of electoral democracies, the faster the rate of agricultural decline in 1965–80, the higher the level of agricultural protection in 1980. This interaction is statistically significant in all six regressions. Sympathy for a rapidly declining farm sector, or effective protest by its members, brings more protection under democratic institutions.

Several real-world contrasts in the regression sample produced this result. Among developing countries, a good illustration is the contrast between Thailand and Côte d'Ivoire. Thailand, an electoral democracy (to some extent) with a rapid shift out of agriculture, imposed less taxation on producers of exportables and of cotton than did Côte d'Ivoire, with its one-party state and stable sectoral shares.

[19]Reverse causation (or simultaneity bias) is another fear about results like those relating to the size of the agricultural sector. Perhaps protection raises the share in agriculture, biasing the coefficient of sector size on protection. But such a bias does not seem to explain why the two coefficients in a quadratic function should be so unstable across regressions. More generally, reverse causations are ignored here in the belief that they have had little effect on the results for the disaggregated commodity categories used in this sample. It is certainly plausible, though, that higher protection could have affected some of the independent variables. Higher protection could have raised the relative income of agriculture per occupied person (Y_a/Y_{non}), raised world price instability, and slowed the rate of agricultural decline.

Supporting the same hypothesis with a different policy result is the contrast between Senegal and other African nations in the sample (Cameroon, Togo, Ghana, and Sudan). Senegal is the only African democracy south of the Sahara in the sample. Yet in Senegal, unlike Thailand, the labor force did not shift out of agriculture between 1965 and 1980. The hypothesis thus predicts no contrast in agricultural policy between Senegal and other African nations, and there is indeed no contrast.

For higher-income nations, the interaction between democracy and rapid agricultural decline helps account for other contrasts. It explains most of the greater support for agriculture in Japan than in the Netherlands, which had a similar income per capita in 1980. Within the EC, the hypothesis helps explain the greater support for agriculture in Italy, where the share of agriculture dropped considerably, than in the United Kingdom. It also helps explain why both more democratic Italy and Portugal treated farmers more favorably than did Yugoslavia. There is a glaring counterexample, however: undemocratic South Korea under Park and Chun switched to agricultural protection, unlike Turkey and Mexico, where the share of agriculture declined at a similar rate.[20] All things considered, the results do recommend considering the interaction of electoral democracy and the shift away from agriculture as influences on agricultural policy, even though, as argued below, this interaction fails to explain the basic developmental shift from taxation to protection.

Relative Incomes of Farmers. Another possible policy determinant is the relative income of the farm sector. Honma and Hayami interpreted this income ratio as a proxy for comparative advantage, noting that it tended to be higher for exportables than for importables. Their interpretation is set aside here and in Table 2.6, where the role of comparative advantage is assigned to a dummy variable representing comparative advantage at zero government intervention. The relative income of agriculture is interpreted as just that, an income ratio, rather than a proxy for comparative advantage.

The regression results in Table 2.6 confirm earlier doubts about whether greater relative poverty of the farm sector would explain why policy favored farmers more in higher-income countries. There is not the slightest hint of a significant link between relative income

---

[20]In Anderson and Hayami (1986), South Korea is also an unexplained outlier in their explanation of agricultural protection. For an alternative explanation, see Timmer (1988).

and agricultural policy. One could object that reverse causation (policy supports raising the relative income of farmers) could obscure a strong negative effect of relative income on policy. A more careful test would deduct the effects of agricultural policy itself from the relative income of the farm sector before regressing policy on relative farm income and other variables. Until such adjusted measures are available in enough cases, however, it can just be noted that the single-equation estimates of Table 2.6 do not suggest any strong negative relationship between the relative (pre-fisc) income of the sector and the support it receives from government.

Commodity-Specific Factors

Not all agricultural price interventions are designed for the general benefit (or detriment) of the agriculture sector as a whole. At least two other possible influences differ greatly by commodity: the degree of price instability in world markets, and the size of efficiency losses due to the misallocation of resources induced by price interventions.

Price Instability. To the extent that protection of agricultural producers is aimed at shielding them against world price shocks, we would expect the higher price instability of some products (for example, sugar, milk, beef, pork) to raise rates of protection for producers in this sector. The regression results support this expectation somewhat. In all six regressions, the average price instability of the commodity emerges as a significantly positive influence on agricultural protection for importable goods. This should be interpreted as part of the larger motif of income sensitivity of farmers to price. Yet there are three limits to the impressiveness of the price-instability effect in Table 2.6. First, it did not extend to the case of exportable goods. Second, the significance of the effect for importables may have been overstated by reverse causation, because heavy protection by many nations can destabilize the price of a product on the world market. Finally, the price-instability effect makes only a negligible contribution to explaining the basic difference between the policies of developed and developing nations.

Deadweight Welfare Losses. The other effect that varies (mainly) by commodity relates to the marginal deadweight costs of price intervention. As sketched in the pressure-group model, the greater the marginal deadweight cost for a given level of intervention, the lower the policy intervention the political marketplace should produce.

Testing this hypothesis calls for considerable behind-the-scenes calculations. The marginal deadweight costs of raising an export tax or an import tariff are first related to price elasticities of exports or imports through standard Marshallian formulas. The foreign trade elasticities are then linked to overall (national or world trade) behavioral elasticities of demand and supply using identities. Then comes the hardest part: finding suitable estimates of basic own-price elasticities. They vary distressingly from study to study, even for the same product and nation. The range of available studies is cited in the Appendix, and median estimates, or "median rumors," are selected for the demand and supply elasticities for each of fourteen products. These are fed into the formulas described in the Appendix. The result is a measure of marginal net national gain per unit of extra gain for agricultural producers due to raising an import duty or lowering an export duty, starting from a 10 percent duty.

The deadweight effects fail to account for broad patterns of policy. Their only look of success is for exportable products, where a greater marginal deadweight gain from helping producers is associated with more observed protection (lower export taxation). In one case the association is statistically significant. Yet the deadweight effects of changes in trade barriers on importable agricultural goods are not significantly positive. The relevant coefficients are generally insignificant, and significantly of the wrong sign in one case. This lack of success admits of two obvious interpretations. One is that the whole idea was wrong, because greater deadweight gains or losses do not weigh into political struggles over agricultural trade policy. Another is that the idea of important deadweight effects is still valid and powerful but that its effect was hidden here by the wrong choice of elasticity measures. The latter, more optimistic, interpretation must face a basic long-avoided question, however: if reliable estimates of price elasticities are elusive after several hundred studies of them, what hope is there for confirming controversial hypotheses whose truth depends on price elasticities?

The Level of State Development

So far the discussion has assumed that the state is a neutral playing field on which various political interests compete. An alternative and more realistic perspective recognizes that agricultural policy can be shaped by the state through efforts to strengthen its own power to govern or to develop the economy. One hypothesis relates to the state as its own pressure group, especially as it seeks revenues to carry out even the minimal functions of government. A second hypothesis, first formulated by Alexander Gerschenkron in the context of European history, argues that more backward states squeeze their agricultural sectors harder in order to finance more rapid industrial revolutions.

Demand for Revenue. As noted previously, the demand of the government for revenue is spurred either by the self-interest of officials themselves or by the loftier infant government argument for trade taxes that support the creation of socially vital state services. Either variant predicts that in such cases governments will tax both exports and imports more heavily than will others with weaker pressures to raise revenue. The revenue-demand argument thus predicts stronger taxation of producers of exportable goods and stronger protection of producers of importable goods. Revenue demand is here proxied by postwar independence, on the reasoning that the postwar nations have a stronger mandate for raising tax revenue from foreign trade.

The results in Table 2.6 are slightly sympathetic to the revenue-demand argument. Certainly they are in accordance with the pattern of policy toward importable goods. In all the relevant regressions, new governments protected agricultural producers against import competition, usually with high statistical significance. And in all regressions, often with statistical significance, producers of importables were more protected than producers of exportables. The only partial setback for the hypothesis is that producers of exportables were not consistently taxed, and in one case, regression (4), they were significantly protected. This aside, demand for revenue by new governments looks like a noteworthy minor actor in the global pattern.

Economic Backwardness and the Gerschenkron Hypothesis. A different kind of state development was hypothesized by Gerschenkron (1962). Nations that resolved to catch up to the economic leaders in the world were more likely than others to tax agriculture to finance modernization. More precisely, Gerschenkron conjectured that the more backward a nation at the time it started a serious attempt to catch up economically, the more it taxes agriculture. To render the hypothesis testable, let us introduce an interaction term equal to the product of the backwardness of the country and its national saving rate, plus another term equal to the saving rate alone. The saving rate represents the seriousness of commitment to accumulation. Both co-efficients should be positive.

For all its intellectual appeal, the Gerschenkron hypothesis fails here. Positive signs predominate, but none is significant, except for the interaction term for developed countries in equation (2), where the saving-rate shift term is significantly negative, contradicting the hypothesis. Whatever Gerschenkron may have seen in the raw corre-lations between catch-up drives and the policy of squeezing agricul-ture, it fails to translate into agricultural trade policies in the global cross section around 1980.

Accounting for the Developmental Pattern

The regression equations, with their decent fit, can also indicate how each force contributes to explaining the observed difference be-tween protection in developed countries and taxation of agriculture in developing countries. One form of accounting is to compare the significance (or the beta coefficients) of the forces just discussed with that of other variables that merely describe the pattern and do not explain it. The merely descriptive variables are those at the bottom of Table 2.6. Here we see shift terms representing the level of GNP per capita in the country and the agricultural product as an exportable good or not. With a couple of exceptions, these variables retell the story we began with: exportables are less protected than im-portables, and higher GNP per capita raises the nominal protection coefficient. The significance of the coefficients for these merely de-scriptive variables means that parts of the basic patterns have eluded all the more causal variables discussed above.

A more frontal accounting appears in Table 2.7. Here each hy-pothesis, armed with its key variables and their regression coeffi-cients, has its chance to account for the observed average difference between the NPCs of the developed and developing countries. The results are stark. Two variables together overpredict the whole dif-ference, and the others either do not matter or predict the opposite of what is observed. The two variables that explain more than 100 per-cent of the difference are sector size and the set of merely descriptive shift terms and income slopes. This might seem like a victory for the sector-size argument, that is, the political potency of small-sector lob-bies. Alas, even this interpretation is weak. Sector size might repre-sent familiar political mechanisms, but it could also be merely descriptive, like the income-related shift terms. It may just be that

Table 2.7. Accounting for the rise of agricultural support with economic development, using regression equations and the 1980 sample

	Contributions of independent variables to difference in average ln(NPC) between 113 DCs and 134 LDCs	
	Using regression (1) in Table 2.6	Using regression (4) in Table 2.6
Independent variable		
Sector size	0.214	0.673
Relative income of agricultural sector	−0.008	−0.051
World price instability for the product	0.030	0.033
Marginal deadweight cost of policy	−0.004	−0.004
New government	−0.092	−0.237
Gerschenkron hypothesis	−0.115	−0.052
Democracy cum agricultural decline	−0.029	0.034
Exportable good	—	0.002
Shift terms and income slopes	0.500	—
Data from Tyers and Anderson	−0.071	−0.062
Predicted difference	0.426	0.337
Actual difference	0.372	0.372

more developed countries have smaller agricultural shares and that the partial correlation to agricultural protection is not in any way caused by sector size. For all its usefulness on several fronts, the regression test cannot unlock the developmental puzzle of agricultural policy because too many of the possible causes are so collinear with development itself. In other words, the available data confine the formal statistical tests to illuminating the byways of global differences in policy but not the main road. For that illumination, we must still rely on the more tentative conjectures of the pressure-group model.

Conclusions

The two patterns in agricultural policy explored here call out for historical perspective. The developmental pattern of today, whereby economic development and rising income per capita bring a switch from taxing agriculture to protecting it, reappears and disappears on the trail back through history. At least from 1860 on, it reappears repeatedly in the long-run history of the now industrialized countries. But in nations that are still building a state apparatus, the common denominator is the anti-trade bias, in which suppliers of importable agricultural products are less taxed or more protected than suppliers of exportables. First in early modern Europe, then again in the rearranged nations of eastern and southern Europe early in this century, and above all in the newly developing countries of the 1960s, an emergent state taxed both imports and exports of agricultural products.

What explains these patterns? The discussion is advanced here by casting new doubt on some explanations, introducing others, and testing as many competing explanations as possible. Several of the leading explanations of why agriculture is protected fall short. Some of them fail to fit the facts, and all share the defect that they explain something that is false—that most governments protect agriculture. Other, more strictly economic explanations are joined logically by the generic pressure-group model.

Combining qualitative observations with formal regression analysis yields a schematic summary of where the issue stands (see Table 2.8). The developmental pattern, in which developing countries tax agriculture and developed ones subsidize it, is probably best explained by a combination of two related forces. First, as the agricultural sector shrinks, farmers increase their effectiveness in lobbying and gain widespread sympathy, especially in electoral democracies in which economic decline of the sector is rapid. Second, the agricultural lobby is also increasingly mobilized by the rising income sensitivity of farm operators and landowners to price movements that government could prevent. Both forces are hard to test directly and formally because they are so closely correlated with economic development itself. There is, however, clear evidence against other possible explanations of the developmental pattern, as listed in Table 2.8.

Why was the shift in the treatment of agriculture more pronounced than the changes in policy toward other sectors? The same two leading candidates emerge, with some minor changes in emphasis. The overall drop in the share of agriculture in the economy was greater than for other sectors, so the power-of-small-lobbies argument could again be applied to the task of explaining what is different about agriculture. It is also likely that the income sensitivity to price is greater for agriculture than for other sectors.

A third possible explanation of why agricultural policy changed more dramatically still awaits a fair test. Perhaps the state could tax agriculture heavily in early settings, and subsidize it heavily in industrialized countries, partly because of the basic price inelasticity of

Table 2.8. Summary of the contributions of competing hypotheses to explaining the net protection or taxation of agriculture

Empirical pattern	Most probable	Less successful	Untested
The shift from LDCs' taxation to DCs' protection	Small sectors lobby powerfully Greater income sensitivity to price	Farm aid is welfare Nostalgia Gerschenkron hypothesis	Income sensitivity to price (as per Anderson, 1987)
This shift was more dramatic for agriculture than for nonagriculture	Small sectors lobby powerfully Greater income sensitivity to price	Same as above	Deadweight cost effects
The anti-trade bias (tax exportables, protect importables)	State's revenue demands	Food security concerns	
Commodity pattern of NPCs within agriculture	Income sensitivity to price	Deadweight cost effects (as per Gardner)	Lobbying role of marketing organizations
Corn Law exception	Price-discrimination model of fiscal exploitation by a socially distant landed elite	Food security concerns	
Role of electoral democracy	Interaction of democracy with agricultural decline raises protection		

Note: In this table, *protection* means income support in any form, including outright subsidies, not just protection against import competition.

agricultural demand and supply. The greater price inelasticity of agriculture should mean that any given amount of government redistribution, whether toward agriculture or at its expense, could be achieved with lower net deadweight costs than in other sectors of the same size and trade orientation. When agriculture is involved, there is less scandalous national waste to dampen the advocates of redistribution and to mobilize its opponents.

The anti-trade bias of agricultural policy seems most easily explained by the demand of the state for revenue. For reasons noted earlier, concerns for food security do not seem to predict the policies actually observed.

As for the commodity pattern of agricultural protection, the regression tests do not support the recent hypothesis that differences in deadweight costs explain which farm products are more subsidized. More likely, the commodity pattern, which is not necessarily robust, is better explained by differences in the sensitivity of the agricultural subsector to world price movements.

We should expect a slow retreat from the heavy taxation of agriculture in developing countries as the infant government excuse for such taxes recedes. But whether they follow the trend in agricultural policy in the industrialized countries may be affected by two clear sources of erosion of their willingness to subsidize farmers. First, agricultural interests lobby more successfully in those electoral democracies in which the share of agriculture in the economy has just shrunk rapidly. In the industrial market economies of today, that shrinkage cannot match the decline in the agricultural labor force that Japan and a few other nations experienced between 1965 and 1980—on the order of 15 percent. Less potential for depopulation of a politically strong sector leaves less sympathy and less power for its lobby. The second source of erosion is the possible exhaustion of the credit-worthiness of central government. In future budgetary showdowns, agricultural subsidies could be decimated, leaving only the import-competing sectors heavily protected, as in Japan.

Acknowledgments

I am deeply indebted to Maite Cabeza-Gutés for research assistance. Helpful criticisms have been received from Julian Alston, Colin A. Carter, C. Peter Timmer, and participants in the Marbach Conference on Agriculture and Economic Development in the 1980s: Lessons for the Role of Government, August 1989, sponsored by the Jacobs Suchard Company, and seminars at California State University–Hayward, the Ecole des Hautes Etudes en Sciences Sociales and Maison des Sciences de l'Homme, Paris, the Greater Chicago Area Group in Economic History, and Iowa State University.

Appendix: Calculating Efficiency Losses
from Price Interventions

Agriculture's share of the labor force is measured so that 6 percent is recorded as 0.06. The log of agriculture's relative income per occupied person uses its share of gainful employment (n_a) and of GNP (y_a) to derive $\ln[y_a \cdot (1 - n_a)/n_a \cdot (1 - y_a)]$.

The world price instability measures are sums of absolute deviations from log trend for 1961–72, from Lancieri (1979), and receive support from the estimates of Scandizzo and Diakosawas (1987). These measures are multiplied by EXPORT (which equals 1 if the good would be an exportable with no government intervention) and then by (1 − EXPORT) to produce separate instability measures for exportables and importables.

The variables DEADEX and DEADIM measure the rates of deadweight gain per unit of producer-surplus gain as policy marginally lowers an export duty from a 10 percent rate or raises an import duty above 10 percent. Both are derived from market-share parameters and from underlying own-price elasticities of supply ($\epsilon \geq 0$) and demand ($\eta \leq 0$) that are assumed to be the same whether the suppliers and demanders are domestic or foreign.

Algebraically, the formulas used are

$$\text{DEADEX} = (\epsilon - \eta\sigma_x) \cdot [t/P + (1 - \sigma_f)/(\eta - \epsilon\sigma_f)] \text{ and}$$
$$\text{DEADIM} = (\eta/\sigma - \epsilon) \cdot [t/P - (1 - \delta)/(\epsilon - \eta\delta)],$$

where t/P = the export or import duty as a percentage of the world price, and the market shares are

δ = world exports to countries other than our country as a share of world exports (of a good we import),

σ = our domestic suppliers' share of domestic demand for a good we import,

σ_f = exports from other countries as a share of world exports, for a good we export, and

σ_x = our domestic demanders' share of domestic supply of a good we export.

The tax rate t/P is standardized at 10 percent. Market shares are derived from FAO data. These shares had to be adjusted to the hypothetical market shares that would have obtained if t/P = 10 percent, using the assumed elasticities.

As the text warns, the basic supply and demand elasticities (ϵ and η) are extremely hard to estimate. Hundreds of attempts relating to agricultural products have been summarized by Askari and Cummings (1976), Adams and Behrman (1982), and Carter and Gardiner (1988). The estimates vary enormously. To portray the "median rumor" for each elasticity, I took the medians of the studies reported by these compilations plus the estimates of Gulbrandson and Lindbeck (1973, p. 256), McCalla, White, and Clayton (1986, pp. 6–21), Tyers and Anderson (in Anderson and Hayami, 1986), Great Britain, Ministry of Agriculture, Fisheries and Food (1987, Tables 5.2 and 5.3), and Gardner (1987, p. 299). Most elasticities are biased downward, but their rankings *might* be fairly portrayed by these median estimates of own-price elasticities within nations:

Demand (η)		Supply (ϵ)	
1. beef	−1.10	1. chicken	2.50
2. chicken	−0.92	2. eggs	1.35
3. coffee	−0.60	3. pork	1.12
4. maize	−0.50	4. cocoa	0.77
5. barley	−0.45	5. beef	0.70
6. sugar	−0.45	6. milk	0.60
7. cocoa	−0.40	7. coffee	0.50
8. cotton	−0.40	8. cotton	0.40
9. pork	−0.34	9. sugar	0.40
10. rice	−0.30	10. wheat	0.40
11. wheat	−0.30	11. tea	0.35
12. milk	−0.28	12. rice	0.30
13. tea	−0.25	13. barley	0.10
14. eggs	−0.20	14. maize	0.10

These were used in the derivation of DEADEX and DEADIM.

To pick up the influence of the revenue needs of new governments, I distinguished governments that gained independence only after World War II. Newness of government is predicted to raise duties on both imports and exports, thus protecting suppliers of import-competing products but taxing suppliers of exportable products. The new-government dummy variable was thus separately interacted with EXPORT and with (1 − EXPORT).

The Gerschenkron hypothesis predicts that agriculture will be taxed more heavily by governments making a determined effort to raise national saving and force rapid growth, the more so the more backward the nation. The two Gerschenkron variables are thus the national saving rate and this times the log of the ratio of U.S. GNP per capita ($8,069) to the GNP per capita of the country in question.

The democracy dummy DEMOC = 1 if the nation had legal opposition parties throughout 1978–80 and held free national elections for legislature or head of state any time between 1973 and 1980, otherwise zero. The data source is Banks et al. (1979–82/83). All the fourteen developed countries in the sample (those with GNP per capita above \$4,000 in 1980) were rated as democracies. So were Bangladesh, Colombia, India, Mexico, Portugal, Thailand, and Turkey. Of the Sub-Saharan nations in the sample, only Senegal was rated a democracy. Argentina, Korea, Pakistan, Philippines, Tunisia, and Yugoslavia were not democracies. Two tough cases given an arbitrary value of DEMOC = 0.5 were Egypt under Sadat and Brazil under modified military rule.

The rate of agricultural decline is the percentage decline in agriculture's share of employment between 1965 and 1980 (e.g., a decline from 33 percent to 19 percent is coded as 0.14).

DCs were, again, nations with a GNP per capita over \$4,000 in 1980.

The TA dummy was added to the regression because the Tyers-Anderson estimates of nominal protection coefficients seemed to be higher than alternative estimates for the same products, same countries, same years.

References

Abel, Wilhelm, 1973. *Crises agraires en Europe (XIIIe–XXe siècle)* [Agrarian crises in Europe: Thirteenth to the twentieth century]. French translation of 1935 edition in German. Paris: Flemmarion.

Adams, F. Gerard, and Jere R. Behrman. 1982. *Commodity Exports and Economic Development.* Lexington, Mass.: Lexington Books.

Alston, Julian, and Colin Carter. 1989. "Causes and Consequences of Farm Policy." Paper prepared for the Sixty-fourth Annual Western Economics Association International Conference, Lake Tahoe, June 18–22. Typescript.

Anderson, Kym. 1987. "Rent-Seeking and Price-Distorting Policies in Rich and Poor Countries." University of Adelaide, Department of Economics, November. Typescript.

Anderson, Kym, and Yujiro Hayami, with associates. 1986. *The Political Economy of Agricultural Protection: East Asia in International Perspective.* London: Allen and Unwin.

Askari, Hossein, and John Thomas Cummings. 1976. *Agricultural Supply Response.* New York: Praeger.

Augé-Laribé, M. 1950. *La politique agricole de la France de 1880 à 1940* [Agricultural policy in France, 1880–1940]. Paris: Presses Universitaires Françaises.

Bairoch, Paul. 1976. "Europe's Gross National Product, 1800–1975." *Journal of European Economic History* 5 (Fall): 273–340.

Balassa, Bela, and associates. 1971. *The Structure of Protection in Developing Countries*. Baltimore: Johns Hopkins University Press.

Bale, Malcolm D., and Ernst Lutz. 1981. "Price Distortions in Agriculture and Their Effects: An International Comparison." *American Journal of Agricultural Economics* 63 (February): 8–22.

Balisacan, Arensio M., and James A. Roumasset. 1987. "Public Choice of Economic Policy: The Growth of Agricultural Protection." *Weltwirtschaftliches Archiv* 123, no. 2:232–48.

Banks, Arthur S., et al. 1979–82/83. *Political Handbook of the World*. New York: McGraw-Hill.

Barnes, Donald G. 1930. *A History of English Corn Laws*. Reprint. 1961. New York: Augustus Kelley.

Bates, Robert H. 1988. "Lessons from History, or the Perfidy of English Exceptionalism and the Significance of Historical France." *World Politics* 40, no. 4:517–41.

Becker, Gary S. 1983. "A Theory of Competition among Pressure Groups for Political Influence." *Quarterly Journal of Economics* 98, no. 3:371–400.

———. 1985. "Public Policies, Pressure Groups, and Dead Weight Costs." *Journal of Public Economics* 28, no. 3:329–47.

Benedict, Murray R. 1953. *Farm Policy in the United States, 1790–1950*. New York: Twentieth Century Fund.

Binswanger, Hans P., and Pasquale L. Scandizzo. 1983. "Patterns of Agricultural Protection." World Bank Discussion Paper ARU 15. Washington D.C.: World Bank. Typescript.

Brady, Dorothy S. 1972. "Consumption and the Style of Life." In Lance E. Davis et al., *American Economic Growth*, pp. 61–92. New York: Harper & Row.

Brigden, J. B., et al. 1929. *The Australian Tariff*. Melbourne: Melbourne University Press.

Carter, Colin A. 1988. "Trade Liberalization in the Grain Markets." *Canadian Journal of Agricultural Economics* 36:633–41.

Carter, Colin A., and Walter H. Gardiner. 1988. *Elasticities in International Agricultural Trade*. Boulder, Colo.: Westview Press.

Delorme, Robert, and Christine André. 1983. *L'état et l'economie . . . en France (1870–1980)* [The state and the economy . . . in France, 1870–1980]. Paris: Editions du Seuil.

Díaz-Alejandro, Carlos F. 1970. *Essays on the Economic History of Argentina*. New Haven: Yale University Press.

Fay, C. R. 1932. *The Corn Laws and Social England.* Cambridge: Cambridge University Press.

Feinstein, C. H. 1972. *National Income, Expenditure and Output of the United Kingdom, 1855–1965.* Cambridge: Cambridge University Press.

Fetter, Frank W., and Henry Chalmers. 1924. *Foreign Import Duties on Wheat, Wheat Flour, Meat and Meat Products.* U.S. Bureau of Foreign and Domestic Commerce, Trade Information Bulletin 233. Survey of World Trade Products, no. 5. Washington, D.C.: U.S. Government Printing Office.

Fogel, Robert W. 1989. "Second Thoughts on the European Escape from Hunger. . . ." NBER Working Paper Series on Historical Factors in Long-Run Growth, no. 1. Cambridge, Mass.: National Bureau of Economic Research.

Funk, W. C. 1914. *What the Farm Contributed Directly to the Farmer's Living.* USDA Farmer's Bulletin 635. Washington, D.C.: U.S. Government Printing Office.

Gardner, Bruce L. 1987. "Causes of U.S. Farm Commodity Programs." *Journal of Political Economy* 95 (April): 290–310.

———. 1988. *The Economics of Agricultural Policies.* New York: Macmillan.

Gerschenkron, Alexander. 1943. *Bread and Democracy in Germany.* Berkeley: University of California Press.

———. 1962. *Economic Backwardness in Historical Perspective.* Cambridge: Belknap Press of Harvard University Press.

Golub, E. O. 1944. *The Méline Tariff: French Agriculture and Nationalist Economic Policy.* New York: Columbia University Press.

Gourevitch, Peter Alexis. 1977. "International Trade, Domestic Coalitions, and Liberty: Comparative Responses to the Crisis of 1873–1896." *Journal of Interdisciplinary History* 8 (Autumn): 281–313.

Great Britain, Board of Trade. 1909. "British Foreign Trade and Industry." In House of Commons, *Sessional Papers,* 102:896–909.

Great Britain, Committee on Industry and Trade. 1927. *Survey of Overseas Markets.* London: Her Majesty's Stationery Office.

Great Britain, Ministry of Agriculture, Fisheries, and Food. 1987. *Household Food Consumption and Expenditure: 1985.* London: Her Majesty's Stationery Office.

Gulbrandsen, Odd, and Assar Lindbeck. 1973. *The Economics of the Agricultural Sector.* Stockholm: Almqvist and Wiksell.

Hayami, Yujiro. 1988. *Japanese Agriculture under Siege: The Political Economy of Agricultural Policies.* New York: St. Martin's Press.

Heckscher, Eli F. 1934. *Mercantilism.* Vols. 1 and 2. London: Allen and Unwin.

Honma, Masayoshi, and Yujiro Hayami. 1986. "The Determinants of Agricultural Protection Levels: An Econometric Analysis." In Kym Anderson and Yujiro

Hayami, with associates, *The Political Economy of Agricultural Protection: East Asia in International Perspective*, pp. 39–49. London: Allen and Unwin.

Ingram, James C. 1971. *Economic Change in Thailand, 1850–1970.* Stanford: Stanford University Press.

Johnson, D. Gale. 1973. *World Agriculture in Disarray.* New York: Macmillan.

Kaplan, Steven L. 1976. *Bread, Politics, and Political Economy in the Reign of Louis XV.* 2 vols. The Hague: Martinus Nijhoff.

Kindleberger, Charles P. 1951. "Group Behavior and International Trade." *Journal of Political Economy* 59 (February): 30–47.

———. 1975. "The Rise of Free Trade in Western Europe, 1820–1975." *Journal of Economic History* 35 (March): 20–55.

Kristov, Lorenzo, Peter Lindert, and Rob McClelland. 1989. "Pressure Groups and Redistribution." Working Papers in Applied Macroeconomics. University of California–Davis, Institute of Governmental Affairs, Davis, Calif.

Lancieri, Elio. 1979. "Instability of Agricultural Exports: World Markets, Developing and Developed Countries." *Banca Nazionale del Lavoro Quarterly Review* 131:287–310.

Laslett, Peter, ed. 1973. *The Earliest Classics: John Graunt . . . Gregory King.* London: Gregg.

Liepmann, Heinrich. 1938. *Tariff Levels and the Economic Unity of Europe.* London: Allen and Unwin.

Lindert, Peter H. 1986. "Unequal English Wealth since 1670." *Journal of Political Economy* 94, no. 6:1127–62.

———. 1987. "Who Owned Victorian England? The Debate over Landed Wealth and Inequality." *Agricultural History* 61 (Fall): 25–51.

———. 1989. "Modern Fiscal Redistribution: A Preliminary Essay." Working Paper 55. University of California–Davis, Agricultural History Center, Davis, Calif.

Little, Ian M. D., Tibor Scitovsky, and Maurice Scott. 1970. *Industry and Trade in Some Developing Countries: A Comparative Study.* London: Oxford University Press for the OECD.

Lucier, Gary, Agnes Chesley, and Mary Ahearn. 1986. *Farm Income Data: A Historical Perspective.* Statistical Bulletin 740. Washington, D.C.: U.S. Department of Agriculture, Economic Research Service.

Lutz, Ernst, and Pasquale L. Scandizzo. 1980. "Price Distortions in Developing Countries: A Bias against Agriculture." *European Review of Agricultural Economics* 7, no. 1:5–27.

McCalla, Alex F. 1969. "Protectionism in International Agricultural Trade, 1850–1968." *Agricultural History* 44 (July): 329–44.

McCalla, Alex F., T. K. White, and K. Clayton. 1986. *Embargoes, Surplus Disposal, and U.S. Agriculture.* USDA Agricultural Economics Report 564. Washington, D.C.: U.S. Government Printing Office.

McGuire, Robert A. 1981. "Economic Causes of Late Nineteenth Century Agrarian Unrest: New Evidence." *Journal of Economic History* 41, no. 4:835–52.

Mayhew, Anne. 1972. "A Reappraisal of the Causes of Farm Protest in the United States, 1870–1900." *Journal of Economic History* 32, no. 2:464–75.

Mitchell, Brian R. 1981. *International Historical Statistics—The Americans and Australasia.* Detroit: Gale Research.

Mitchell, Brian R., with the collaboration of P. Deane. 1962. *Abstract of British Historical Statistics.* Cambridge: Cambridge University Press.

Olson, Mancur. 1965. *The Logic of Collective Action.* Cambridge: Harvard University Press.

——. 1985. "Space, Agriculture, and Organization." *American Journal of Agricultural Economics* 67 (December): 928–37.

Organization for Economic Cooperation and Development (OECD). 1987. *National Policies and Agricultural Trade.* 8 vols. Paris.

Outhwaite, R. B. 1981. "Dearth and Government Intervention in English Grain Markets: 1590–1700." *Economic History Review,* 2d ser., 34:380–406.

Petit, Michel. 1985. *Determinants of Agricultural Policies in the United States and the European Community.* IFPRI Research Report 51. Washington, D.C.: International Food Policy Research Institute.

Posthumus, N. W. 1946. *Inquiry into the History of Prices in Holland.* Leiden: E. J. Brill.

Ranis, Gustav. 1959. "The Financing of Japanese Economic Development." *Economic History Review,* 2d ser., 11 (April): 440–54.

Scandizzo, Pasquale L., and Dimitris Diakosawas. 1987. *Instability in the Terms of Trade of Primary Commodities, 1900–1982.* FAO Economic and Social Development Paper 64. Rome: Food and Agriculture Organization of the United Nations.

Schiff, Maurice, and Alberto Valdés, eds. Forthcoming. *The Economics of Agricultural Price Interventions in Developing Countries.* Baltimore: Johns Hopkins University Press.

Schultz, Theodore W. 1953. *The Economic Organization of Agriculture.* New York: McGraw-Hill.

——. 1978. *Distortions of Agricultural Incentives.* Bloomington: Indiana University Press..

Smith, Thomas C. 1969. "Farm Family By-Employments in Preindustrial Japan." *Journal of Economic History* 29 (December): 687–715.

Spring, David, ed. 1977. *European Landed Elites of the Nineteenth Century.* Baltimore: Johns Hopkins University Press.

Timmer, C. Peter. 1988. Review of *The Political Economy of Agricultural Protection: East Asia in International Perspective* by Kym Anderson and Yujiro Hayami, in *Asian-Pacific Economic Literature* (Canberra) 2 (September): 66–69.

Timmer, C. Peter, Walter P. Falcon, and Scott R. Pearson. 1983. *Food Policy Analysis.* Baltimore: Johns Hopkins University Press for the World Bank.

Tracy, Michael. 1982. *Agriculture in Western Europe: Challenge and Response, 1880–1980.* 2d ed. London: Granada.

Tyers, Rodney, and Kym Anderson. 1986. "The Price, Trade and Welfare Effects of Agricultural Protection." In Kym Anderson and Yujiro Hayami, with associates, *The Political Economy of Agricultural Protection: East Asia in International Perspective,* pp. 50–62. London: Allen and Unwin.

——. Forthcoming. *Distortions in World Food Markets.* Cambridge: Cambridge University Press for the Trade Policy Research Centre.

U.S. Census Bureau. 1976. *Historical Statistics of the United States from Colonial Times to 1970.* Washington, D.C.: U.S. Government Printing Office.

U.S. Department of Agriculture (USDA), Economic Research Service. 1988. *Estimates of Producer and Consumer Subsidy Equivalents: Government Intervention in Agriculture, 1982–1986.* Washington, D.C.

U.S. Secretary of Agriculture. 1933. *World Trade Barriers in Relation to American Agriculture.* 73d Cong., 1st sess, Senate Document 70. Washington, D.C.: U.S. Government Printing Office.

U.S. Tariff Commission. 1934. *The Tariff and Its History.* Washington, D.C.: U.S. Government Printing Office.

Wilson, Constance M. 1983. *Thailand: A Handbook of Historical Statistics.* Boston: G. K. Hall.

World Bank. 1982. *World Development Report, 1982.* New York: Oxford University Press for the World Bank.

——. 1985. *World Development Report, 1985.* New York: Oxford University Press for the World Bank.

——. 1986. *World Development Report, 1986.* New York: Oxford University Press for the World Bank.

Wright, Philip G. 1935. *Trade and Trade Barriers in the Pacific.* Stanford: Stanford University Press.

3

The Role of Agricultural
Exports in Development

Alberto Valdés

The role of agricultural exports in development is to a large extent conditioned by the overall trade strategy of a particular country rather than by its agricultural strategy. Furthermore, while export prospects could be improved by a more liberal import policy on the part of the rich industrial countries, a developing country's domestic policies are of special significance if it is to take advantage of the opportunities offered by foreign trade to stimulate economic development. One of the lessons learned from the various trade strategies for the industrial sector since the 1960s is that they make a greater contribution to overall economic growth when export-oriented strategies are followed. Although we do not yet have a systematic empirical analysis of agriculture as we have for the industrial sector, agricultural export-led growth has real potential in a variety of settings.

The development strategies after World War II in most developing countries (LDCs) grossly undervalued the potential contribution of agriculture in general, and particularly of agricultural exports, to economic development. Governments and policy analysts in developing countries have now begun to reassess the potential contribution of agriculture as a strategic element of the development process. As part of this strategy, agricultural exports can play a critical role in stimulating agricultural growth, indirectly generating rural employment and hence alleviating poverty, in addition to contributing directly to foreign exchange earnings. The extraordinarily high taxation of the production of agricultural exportables in most developing countries means that there is considerable scope for fostering incentives to expand their production of exportables.

The first section of this chapter presents a condensed version of the prevailing view in the 1950s and 1960s about the limited role that expanded exports of primary products could play in stimulating domestic development. At the time, this was essentially a theoretical debate. But theory does not define the relative importance of exporting and import-competing activities in an optimal allocation of resources and the effect of that allocation on the rate of economic growth. More recently, based on the contrasting experience of various developing countries during the last thirty-five years, empirical analysis has shown a strong association between export performance and growth. Disillusionment with the performance of inward-oriented development strategies led to their reassessment; the positive role of outward-oriented strategies has been increasingly recognized. Since the agricultural sector in most developing countries is a source of a wide range of tradable goods, agriculture can play a more active role under a neutral trade and exchange-rate regime that does not discriminate against exports.

The second section of the chapter describes the trends, structure, and performance of agricultural exports of developing countries during 1962–84. In many LDCs, particularly in Sub-Saharan Africa, agricultural export performance, based on market share criteria, has been considerably below its potential. The reasons for this lagging performance are examined in the third section, which argues that agricultural exports were taxed heavily, partly as a means of generating government revenues but mainly as an indirect and perhaps unintended effect of economywide policies, an inherent consequence of the development strategies followed since World War II. The final section discusses specific current issues and concerns about the adoption of outward-oriented strategies for agriculture.

The Contribution of Agricultural Exports to Development: The Debate

There was never a question that exports from developing countries were dangerously large. More often the concern was that the volume of exports was not sufficiently large to provide enough foreign exchange to finance new investment. Moreover, it was not surprising that exports of developing countries originated from a narrow range of subsectors within the economy, usually with a heavy reliance on natural resources. Economists were pessimistic about the growth of external demand for exports of primary products, however, and skeptical about the dynamic influence of exports on the rest of the economy. Given the perceived risk of specializing in the export of a narrow range of primary products, they were also skeptical about whether LDCs could rely on a strategy whereby economic growth is induced from the outside through an expansion of world demand for primary products, particulary in the case of exports from tropical latitudes, which were assumed to face inelastic demand from industrial countries with respect to price and income (see, for example, Nurske, 1961). More broadly perhaps, it was perceived that the import capacity of the country rode on cycles of boom and bust, depending on the prices its agricultural exports could reap in distant markets. This inability of a country to control its own economic destiny made policy makers apprehensive and alienated them from outward-oriented development strategies.

Such a pessimistic perception was widespread, even though some economists in the 1950s and 1960s were arguing that in selecting a development strategy, trade should be considered as more than an exchange of goods, particularly between economies at different stages of development. Cairncross (1962, p. 214) argued, for example, "As often as not, it is trade that gives birth to the urge to develop, the knowledge and experience that make development possible, and the means to accomplish it." An outward-looking orientation enables an economy to make more flexible economic adjustments, adjust better to external shocks, and bring the economy closer to an optimal allocation of resources.

In the debate, far too much emphasis was put on the forces operating to limit the demand for primary products and far too little on the effect of the continuous opening up of new markets and new products (see Myint, 1955, 1958). An agricultural export orientation was often associated with a continuation of colonial patterns—a country remained highly dependent on a very narrow range of exports and could expect only a slow rate of growth, because it depended on the export of products subject to great variability in supply, with low elasticity of supply. Even in cases of substantial growth in exports, there was a general perception that exports, including those from agriculture, had not acted as a key sector to stimulate growth in the rest of the economy; the sector could have no sustained and widespread effect because of its weak links with the rest of the economy.[1] This pessimistic attitude about agriculture dominated the 1950s and 1960s and was articulated, for example, by Hirschman (1958), who condemned the sector for its lack of linkages.[2] That trade did not have a more stimulative effect was attributed to a host of domestic impediments that limits the transmission of the gains from exports to other sectors. The determinants of this weak carryover from exports were thought to be fragmented domestic markets, price rigidity, factor immobility, and ignorance of the technological possibilities (Cairncross, 1962; Meier, 1975).

As a result of these perceptions, the attention of policy analysis in the 1950s and 1960s was focused on removing the domestic impediments that limit the transmission of the production of various exports. With regard to removing domestic impediments, the analysis pointed to the need for such measures as the following: increased investment in transportation, communication, and education; policies to diminish the prevalence of monopolistic practices; efforts to widen capital markets; and removal of restraints on land tenure and land use, as well as other factors that would facilitate a shift in resources to export products with a rising demand and thus in general allow a country to take full advantage of new export opportunities.

With regard to the technology required, various exports will have differential direct effects on such factors as the spread of technical knowledge, the demand for skilled labor, the acquisition of organizational and entrepreneurial skills, and the extent to which they stimulate development of new methods of production. Traditionally, a higher degree of processing of agricultural exports was associated with promoting these dynamic gains and stimulating backward and forward linkage (Hirschman, 1958).

Some exports have more dynamic linkages than others. In spite of Hirschman's condemnation of agriculture, little was known from empirical research about the performance of agricultural development in general, and agricultural exports in particular, and their contribution to domestic employment and income generation. The magnitude of such impact would depend on the strength of the linkage or multiplier effects, which in turn is determined by the degree of

---

[1]That is, through its effects on input-supplying industries elsewhere in the economy (backward linkages) and on the development of processing activities based on the output of the export industry (forward linkages) and the linkages derived from consumption expenditures by farmers and workers in export activities.

[2]See Timmer (1988) for a comprehensive analysis of the debate on the role of agriculture in the various stages of development.

infrastructure development and, among other things, by the expenditure patterns of the beneficiaries of the increased income from exports (Mellor, 1976).

Judging from the postwar experience of LDCs, there is now considerable agreement that overall growth performance has been better under export-oriented strategies than under import-substitution strategies. The success stories are well known. They include Côte d'Ivoire in the 1970s, South Korea, Taiwan, Brazil, Malaysia, Thailand, Colombia, and Turkey. By contrast, such inward-looking countries as India, Egypt, and Argentina have experienced sluggish growth. Compared with an import-substitution strategy, export orientation has been shown to create more employment. Outward-looking economies have also been better able than inward-looking ones to adjust to external shocks. More generally, export-oriented policies bring the economy closer to an optimal allocation of resources.[3] As argued by Krueger (1978), the commitment to an export-oriented development strategy implies a fairly liberal and efficient trade regime; rent-seeking behavior, both of entrepreneurs and of government officials, is constrained. They simultaneously receive feedback about the success or failure of policies and investments. Although most of the more recent analysis of the experience of developing countries described above refers to the industrial sector, there is no strong reason to believe that those general attributes of export-oriented policies would not apply to agriculture. Analysis of the experience with various agricultural strategies is not sufficiently advanced, however, to draw conclusions that would apply to many developing countries, particularly results that would indicate the relative importance of agricultural exports for rural employment and for multiplier effects beyond the agricultural sector—the "propulsive" effect of agriculture on the overall economy.

A Quantitative Overview of Agricultural Exports

A brief description of the trends and structure of agricultural exports from developing countries since the early 1960s can serve as background for examining the role of exports in development. First,

[3]For the impact of an export-oriented strategy on overall growth, employment, adjustment to external shocks, and resource allocation, see, respectively, Krueger (1980) and (1978), Balassa (1984), and Feder (1982).

Table 3.1. Commodity structure of agricultural exports from developing countries, 1962–65 to 1981–84

| Commodity | Agricultural exports (million 1984 U.S. dollars) | | | | |
	1962–65	1966–70	1971–75	1976–80	1981–84
Coffee	7,965	8,784	8,064	12,234	7,729
Animal feed	1,968	2,433	2,556	3,026	3,759
Sugar and honey	2,250	2,372	6,230	3,680	3,114
Crude rubber	3,374	2,930	2,329	2,725	2,272
Cotton	5,175	4,999	4,646	3,438	2,244
Fruit, fresh	1,740	2,023	1,971	2,097	2,185
Other vegetable oils and fats	1,227	1,209	1,905	2,336	2,004
Rice	1,594	1,499	1,272	1,797	1,781
Cocoa	2,862	3,594	3,513	4,844	1,668
Vegetables, fresh	829	1,108	1,355	1,780	1,506
Meat, fresh	1,504	1,815	2,083	1,403	1,459
Maize	1,215	1,781	1,794	1,228	1,398
Tea	2,345	1,887	1,304	1,334	1,317
Oilseeds, nuts, and kernels	2,601	2,292	2,370	2,078	1,293
Vegetable oils and fats	715	737	1,016	1,193	1,246
Tobacco, unmanufactured	638	707	999	1,096	1,226
Wheat and meslin	1,269	1,083	760	851	1,206
Fruit, preserved	262	407	559	847	1,117
Cereals, others	238	435	607	632	719
Crude vegetable	536	570	632	666	676
TOTAL	40,305	42,664	45,966	49,285	39,918

Source: Author's calculations.
Note: Deflated by UNCTAD import deflator. Principal agricultural products exported by LDCs ranked according to their value in 1981–84. Numbers represent average annual values over the specified subperiod. The commodity structure in this table is presented at two-digit Standard International Trade Classification (SITC) aggregation.

three questions are addressed. What were the principal products exported since the early 1960s? What was the rate of growth in agricultural exports from LDCs? What was the destination of exports from LDCs by commodity group? The chapter then looks at the performance of developing countries and identifies the factors that contributed to the success or failure of export expansion.

Trends and Structure of Agricultural Exports from Developing Countries

Table 3.1 presents the commodity structure of agricultural exports from developing countries during the period 1962–84. Coffee was the single most important export commodity throughout the period, but exports with the highest value, in additon to coffee, during the 1981–84 period included such commodities as livestock feed, sugar, rubber, cotton, fresh fruits, and vegetable oils.[4] Of these exports there is a large number of tropical products for which there is only limited competition from developed countries; these exports are thus governed closely by the level of world demand and much less so by their ability to compete for market shares vis-à-vis nontropical products. It can also be observed in Table 3.1 that exports of cereals and livestock products are a comparatively small proportion of total agricultual exports from LDCs.

The rate of growth in the value of livestock feed, tobacco, fruits, and fresh vegetables in international trade is remarkably high. The recent rapid growth of nontraditional exports such as horticultural products from some developing countries is outstanding. Since 1984, for example, Kenya and Chile have registered an average growth of more than 20 percent a year in the value of horticultural exports. By contrast, the value of traditional products exported from LDCs, such as tea, rice, cotton, and rubber, remained stable or fell during 1962–84.

Table 3.2 presents the destination of agricultural exports from developing countries by commodity groups during the 1981–84 period. In percentage terms, LDCs are principal export markets for other LDCs for only sugar, rice, meat, maize, tea, wheat, and vegetable oils and fats. The member countries of the Organization for Economic Cooperation and Development are by far the principal market for most of the agricultural exports from developing countries. In terms of the value of LDC agricultural exports in U.S. dollars, it is evident in Table 3.2 that in 1981–84 approximately 59 percent went to the OECD markets, approximately 24 percent went to other LDCs, and the rest went to centrally planned counties. This distribution shows greater diversification than that found in 1962–64, when only 12 percent of total LDC agricultural exports were traded among developing countries, 10 percent were shipped to centrally planned countries, and 78 percent went to the developed countries. Considering the higher value added of processed agricultural exports, an important question for future research is to determine which of the three groups of countries is the principal market for these exports from developing countries. The still high concentration of LDC exports to

---

[4]Livestock feed includes oilseed cake and meal, fish meal, and vegetable products used to feed livestock.

Table 3.2. Destination of exports from developing countries, 1981–84

Commodity	Destination of exports (percent)			Total value of world exports (million U.S. dollars)
	OECD	LDC	CP	
Coffee	87	6	7	7,729
Animal feed	64	12	24	3,759
Sugar and honey	43	39	18	3,114
Crude rubber	68	12	20	2,272
Cotton	51	18	31	2,244
Fruit, fresh	63	24	13	2,185
Other vegetable oils and fats	58	31	12	2,004
Rice	12	77	11	1,781
Cocoa	79	4	17	1,668
Vegetables, fresh	76	20	4	1,506
Meat, fresh	50	41	9	1,459
Maize	14	42	44	1,398
Tea	40	44	16	1,317
Oilseeds, nuts and kernels	55	25	20	1,293
Vegetable oils and fats	24	61	15	1,246
Tobacco, unmanufactured	74	12	14	1,226
Wheat and meslin	2	38	60	1,206
Fruit, preserved	91	7	2	1,117
Cereals, other	29	28	44	719
Crude vegetable	79	18	2	676
Wool and animal	56	12	32	592
Total value of imports (million U.S. dollars)	23,857	9,617	7,058	40,511

Source: Author's calculations.
Note: Deflated by UNCTAD import deflator. These commodities are ranked according to their value in the last column, Total value of world exports. Numbers represent average annual values over the specified subperiod. The commodity structure of this table is presented at two-digit SITC aggregation. CP refers to centrally planned economies.

OECD member countries reinforces the importance of policy reform within OECD to achieve a more liberal trade regime to allow a faster expansion of agricultural exports from developing countries.

Past Performance of Agricultural Exports

Except for cereals, developing countries have been major suppliers of the most important agricultural commodities in world markets. Furthermore, not only do agricultural exports constitute the major source of foreign exchange for many LDCs, but their production also represents a large economic sector in most of them. This section highlights some important findings from studies that examined the performance of the major agricultural export commodities in developing countries since the early 1960s.

Market share analysis is a simple way of differentiating countries according to their export performance and thus identifying various factors involved. A decreasing market share for a country's major agricultural exports is an indication that external demand was not a dominant constraint, and so for a country that experienced a decline in its share, one should look for domestic factors that caused the poor export performance.

As expected, the analysis of annual rates of growth in agricultural exports among LDCs shows a wide variation, reflecting in part the different mix of commodities exported by various countries, difference among products in constraints on external demand, the varied strength of the export sector, and differences in policies for exchange rates and agricultural and nonagricultural trade.

An analysis of market share for 1971–74 relative to 1961–65 found that a sample of twenty developing countries could be classified into three groups according to their performance in exporting agricultural products (Valdés and Huddleston, 1977). The first group of countries—including Turkey, Brazil, and Thailand—had highly diversified agricultural exports and had performed quite well in export markets. For most of their important export products, the gains in market share were substantial. In a second category, countries had managed to increase their market share in agricultural exports in spite of a high export concentration. This group included Cameroon, the Philippines, Malaysia, Sierra Leone, Indonesia, and a few others.

At the other end of the spectrum was a third and large group of countries—including Sri Lanka, Ghana, Bangladesh, Ecuador, Tanzania, Argentina, and Pakistan—for which the performance of their principal export commodities was below the world average. For example, the export volume of each of the commodities from Sri Lanka declined in absolute terms or increased at a lower rate than the world average rate of growth of export of the same commodity. Similarly, Ghana experienced a negative rate of growth in its earnings from its principal export, cocoa beans; its export earnings from cocoa were 53 percent below what they would have been had it maintained its share in the world market.

Although there are several relatively successful cases of vigorous growth in agricultural exports, many LDCs suffered a severe decrease

Table 3.3. Export market shares of cocoa and palm oil in selected developing countries, 1961–63 and 1982–84

Region and country	Shares of export market (percent)	
	1961–63	1982–84
Cocoa		
Africa	80.0	64.1
Cameroon	6.8	6.9
Côte d'Ivoire	9.3	26.3
Ghana	40.1	14.4
Nigeria	18.0	11.2
Latin America	16.7	18.5
Brazil	7.3	10.9
Ecuador	3.2	2.6
Palm Oil		
Africa	55.8	1.9
Nigeria	23.3	0.2
Zaire	25.1	0.1
Asia	41.8	95.0
Indonesia	18.4	8.2
Malaysia	17.9	70.6

Source: Data from Alasdair MacBean, "Agricultural Exports of Developing Countries: Market Conditions and National Policies," in Nurul Islam, ed., *The Balance between Industry and Agriculture in Economic Development* (London: Macmillan, 1989), pp. 101–28.

either in the volume of their principal exports, in the market share, or both. This decline in market share suggests that an unfavorable export performance from the 1961–65 period to 1971–74 seems to be caused more by a country's own domestic policies than by a depressed world market. These trends occurred during a period of vigorous economic growth and rising world GNP, which provided an economic environment very appropriate for an outward-oriented policy.

In a more recent analysis, MacBean (1989) provides a striking illustration of the poor performance in exporting traditional tropical products shown by several developing countries, particularly in Sub-Saharan Africa (except Côte d'Ivoire). Equally striking is the successful story of palm oil in Malaysia (see Table 3.3). While Ghana lost a substantial share of the cocoa market, as did Nigeria and Zaire in the palm oil market, Malaysia increased its share of the palm oil market by more than threefold.

Although MacBean does not address the temporal pattern of export growth of several LDCs that promoted agricultural exports in the past, this topic should receive attention. The Taiwanese experience suggests that the structure of exports may undergo various stages—for example, from land-intensive to land-saving and processed agricultural exports and then to labor-intensive manufactured exports. The role of agricultural exports in the development process might differ, depending on the country's resource endowment, labor costs, and level of development.

Development Strategies: Agricultural Trade and Industrialization

Between World War II and approximately 1973, real economic growth of the world economy far exceeded that of any other quarter-century. In developed countries, a major contributor to this growth was the rapid liberalization of the international trading system in the industrial sector and greater financial integration facilitated by the Bretton Woods institutions. The pace of this growth diminished in the 1970s and early 1980s. An important point made by Krueger (1988), who has examined this evolution, is that trade liberalization accelerated world economic growth and this growth created an environment that made trade liberalization possible. Although a healthy economic environment was vital for the prospects of outward-oriented policies, the biggest gains from the growing international economy accrued to those countries whose own policies were conducive to growth and reliance on the international market.

In contrast to developed countries in the postwar era, most developing countries adopted trade policies that greatly reduced the ties between their domestic markets and the international economy. Some pursued a strategy of deliberate industrialization aimed at a structural transformation of the economy which favored the industrial sector. What led to this strategy was a combination of circumstances in world markets (the Depression of the 1930s and World War II), economic ideology, and a general belief in a greater role for government. These factors coincided with the conventional wisdom of reducing reliance on export markets for agricultural products—in the process generating government revenues from agricultural export taxes and the operation of parastatals—and of promoting industrialization.

Bias of Government Policy against Agricultural Trade

Early development efforts followed a clear strategy. The resources generated by agriculture would be used to promote industrialization.

Until recently it was not known how badly government policies that implemented this strategy, especially the heavy protection of industry from foreign competition, hurt agriculture. In addition, the direct policy measures used to extract resources from agriculture—direct taxes and indirect taxes through biases in the terms of trade—further discriminated against agriculture for the benefit of industry. Little concern was expressed at the time because resources were not thought to be needed in agriculture; it was industry that lagged so badly behind Western standards.

Given this state of mind about the relative merits of agriculture and industry, a simply taxonomy of the rationale given for protecting industry can be constructed.[5]

Taxes on trade are levied to help finance government and provide balance-of-payment support. The revenue motive would argue for a uniform across-the-board treatment of all items, but, in practice, taxes on trade in most countries had an extremely discriminatory effect on individual activities. The protection motive was thus superimposed on the revenue motive.

The *optimum tariff argument* provides the rationale for shifting the commodity terms of trade in favor of a large exporter through an export tax. Although analytically valid, it applies only as an exception in a few products and in a few countries and even then for relatively short period before substitutes are found.

The *infant industry argument* legitimized policies and trade instruments that protected nascent domestic industry. In much of the literature, agriculture was assumed to be technologically static, whereas industry was supposed to be dynamic. The Green Revolution, the widespread adoption of high-yielding varieties of cereals, in many LDCs has demonstrated the fallacy of this assumption.

Industrial protection is advocated on the grounds that it will lead to a net increase in employment. Paradoxically, much of the recent empirical work on LDCs has shown the opposite—that industrial protection leads to a net decrease in the demand for labor (Little, Scitovsky, and Scott, 1970; Krueger et al., 1981). Unfortunately, empirical work on the impact of the trade regime on agricultural employment has been scarce. Nor is it clear whether industry is, on the whole, more or less labor-intensive than agriculture.[6]

[5]This taxonomy draws on Valdés and Siamwalla (1988).
[6]The evidence presented by Timmer in Chapter 5 suggests that in South and Southeast Asia agricultural GDP is more employment intensive than industry on average and at the margin, which reflects lower labor productivity in the agricultural sector. The pattern for the service sector is more mixed.

Policy interventions are needed to deal with economic instability. There are two strands to this argument. One, specific to agriculture, is that short-run instability in world commodity markets makes them unreliable guides for planning imports and long-run domestic production. If the argument is correct, many LDCs should reduce their reliance on world markets by setting the domestic prices of imported staples higher than world prices. Most LDCs have done so.[7] This strategy raises production and lowers consumption, thus reducing imports, but agricultural exportables are taxed via appreciation of the real exchange rate. Relative to import-substituting food staples, agricultural exportables are doubly discriminated against: they receive no direct protection and are indirectly taxed by the exchange rate.

The second argument is the long-debated relationship between export orientaton, instability in export proceeds, and economic growth.[8] Efforts to stabilize domestic prices do not preclude adoption of a pricing policy based on long-run trends in world markets; the two are analytically distinct and can be separated in practice.[9]

Export pessimism. Whether developing countries can benefit from international trade has been a dynamic area of policy debate since the 1960s. Four interrelated arguments that favor a closing up of the economy in developing countries have been used. First was the now classic argument raised by Singer and Prebisch that the terms of trade move against the tradable products of poor countries. Second, it was argued that if a country expands exports of traditional agricultural commodities, it will find overseas markets limited, and so the increased output may result in falling export revenues.[10] Third, the international market may be limited by the protectionist policies adopted by the importing rich countries. Finally, the "new" export pessimism raises the risk of falling export revenues if many developing countries simultaneously expand exports of traditional agricultural products with low price elasticities of import demand.[11]

As Valdés and Siamwalla (1988) argued, the extremely aggregative nature of these arguments is their chief attraction but also their main weakness. Much of the discussion on price movements overlooks cost-reducing innovations. Also, much of the empirical evidence for a secular fall in the terms of trade is period-specific. Furthermore, the rapid expansion of nontraditional agricultural exports, which has

[7]See the results of the study by Krueger, Schiff, and Valdés (1988) discussed below.
[8]These topics were analyzed by MacBean (1966), for example.
[9]See Timmer (1989) for the rationale for such an approach.
[10]This argument has recently been reexamined by Islam and Subramanian (1989) as it applies to traditional tropical commodities such as tea, coffee, cocoa, and bananas.
[11]See the discussion in Koester, Schafer, and Valdés (1989).

followed outward-oriented trade strategies, has led to greater diversification of exports in these countries. Examples are soybeans and frozen concentrated orange juice from Brazil, fresh fruits and vegetables from Chile, horticultural exports from Kenya, and pineapples and cassava pellets from Thailand. Some traditional exports, such as palm oil from Malaysia and basmati rice from Pakistan, have also expanded. Most countries thus are able both to expand and to diversify agricultural exports in spite of the realities of the external constraints on access to markets. There are perhaps some economies with highly specialized resources, however, that have little flexibility and little choice but to continue a very specialized pattern of exports.

It cannot be denied that inelastic growth in demand and the subsidies and protection in developed countries do constrain expansion of exports from developing countries. But the main thrust of the empirical evidence reviewed here is that for most LDCs, successful export performance is more a function of domestic policies than of external demand, as implied by their loss of market share in several agricultural markets and the inherent anti-export bias of the development strategies in most developing countries.

Agricultural Exports: Extent of the Bias

Until recently, economists gave scant attention to the significance of trade and macroeconomic policies in shaping the economic opportunities faced by agricultural producers. The effects of these policies on agriculture have been neglected partly because of the narrow sectoral orientation of most agricultural policy analysis and partly because of the widespread misconception about the limited role of agriculture in economic development.

In promoting industry, government policies distorted incentives against the production of tradable agricultural goods. An import-substitution strategy leads to an overvalued domestic currency (relative to an equilibrium exchange rate at lower levels of protection), which in turn leads to stagnant production of exports and encourages nonprotected importables. This leads to a foreign exchange crisis, and the solution has generally caused the exchange rate to be distorted even further. The resulting penalty on agriculture is inherent and will last as long as industry is protected; it cannot be eliminated by better management in other areas of economic policy. Protecting industry thus means discriminating against agriculture. Similarly, heavy dependence on foreign assistance or "excessive" levels of absorption (in most cases related to public spending) from booming oil and other mineral exports directly affect the real exchange rate and the profitability of farming. Consequently, the macroeconomic management of the economy—that is, nominal exchange rates, interest rates, wages, international capital flows, and fiscal policy—is of utmost importance to the agricultural sector.[12] The key link is what happens to the real exchange rate. In most LDCs, agriculture is a more highly tradable sector than industry or services. Accordingly, the agricultural sector is best served if the real value of the domestic currency is relatively low (high real exchange rate) in contrast to the widely prevailing appreciation of the real rate in most developing countries.

Although the extent of discrimination against agriculture has not been quantified until very recently, observers of the development process have long been aware that developing countries directly intervene systematically and extensively in pricing of agricultural commodities. But the extent to which economywide interventions dominate the sectoral interventions involving direct price policies was not well recognized.

Krueger, Schiff, and Valdés (1988) report a set of estimates on agricultural price interventions for eighteen developing countries during the 1960–84 period. This project was unique not only for the large sample of countries and commodities covered over a period of twenty-five years but also because it allowed comparability across countries. Applying a common methodology, authors of the country studies estimated the direct price interventions—defined as the nominal and effective rates of protection with no adjustment for the exchange rate or for nonfarm prices—and the indirect effects of economywide policies. There were three major elements in the calculations of the indirect effects: the depreciation of the exchange rate required to eliminate the nonsustainable part of the current account deficit; the depreciation of the real exchange rate due to removal of trade interventions which mainly protect industry; and the change in the price of nonagricultural tradables, also due to the removal of these trade interventions. According to the results, industrial protection has had a greater impact on agricultural incentives than has the current account imbalance. In many cases, industrial protection was so high that it was the hypothetical removal of protectionist policies,

[12]The basic mechanisms are explained in Chapter 5 of Timmer, Falcon, and Pearson (1983) and Lewis (1989). Empirical evidence is shown in Bautista and Valdés (forthcoming).

Table 3.4. Direct, indirect, and total nominal rates of protection for exported products, 1975–79 and 1980–84

| Country | Product | Nominal rates of protection (percent) | | | | | |
| | | 1975–79 | | | 1980–84 | | |
		Direct	Indirect	Total	Direct	Indirect	Total
Argentina	wheat	−25	−16	−41	−13	−37	−50
Brazil	soybeans	−8	−32	−40	−19	−14	−33
Chile	grapes	1	22	23	0	−7	−7
Colombia	coffee	−7	−25	−32	−5	−34	−39
Côte d'Ivoire	cocoa	−31	−33	−64	−21	−26	−47
Dominican Republic	coffee	−15	−18	−33	−32	−19	−51
Egypt	cotton	−36	−18	−54	−22	−14	−36
Ghana	cocoa	26	−66	−40	34	−89	−55
Malaysia	rubber	−25	−4	−29	−18	−10	−28
Pakistan	cotton	−12	−48	−60	−7	−35	−42
Philippines	copra	−11	−27	−38	−26	−28	−54
Portugal	tomatoes	17	−5	12	17	−13	4
Sri Lanka	rubber	−29	−35	−64	−31	−31	−62
Thailand	rice	−28	−15	−43	−15	−19	−34
Turkey	tobacco	2	−40	−38	−28	−35	−63
Zambia	tobacco	1	−42	−41	7	−57	−50
AVERAGE		−11	−25	−36	−11	−29	−40

Source: Anne O. Krueger, Maurice Schiff, and Alberto Valdés, "Agricultural Incentives in Developing Countries: Measuring the Effect of Sectoral and Economywide Policies," *World Bank Economic Review* 2, no. 3 (1988): 255–72. Reprinted by permission of the World Bank.

Note: Korea and Morocco are not included because all main agricultural products are imported. The direct nominal rate of protection is defined as the difference between the total and the indirect nominal rates of protection, or equivalently, as the ratio of (1) the difference between the relative producer price and the relative border price, and (2) the relative adjusted border price measured at the equilibrium exchange rate and in the absence of all trade policies.

which led to the decline in the prices of nonagricultural tradables relative to agricultural prices, that dominated the indirect effects. Since industrial protection acts both through the exchange rate and through the relative prices of tradables, it has a large negative effect on agriculture in developing countries.

The direct, indirect, and total nominal protection rates were estimated for representative agricultural export crops of the eighteen countries (see Table 3.4). The figures on direct intervention are estimates of the percentage by which domestic producer prices diverge from those that would have prevailed in the absence of government interventions (at the actual exchange rate and degree of industrial protection). This measure is equivalent to the rate of nominal protection. It is evident that most countries adopted policies that resulted in the equivalent of export taxes. Exceptions were Chile, Zambia, Turkey from 1975 to 1979, and Ghana (where indirect effects are very high). The reduction in export prices in 1975–79 equaled or exceeded 25 percent in Argentina, Côte d'Ivoire, Egypt, Malaysia, Sri Lanka, and Thailand. For 1980–84, all countries except Ghana, Portugal, and Zambia had negative direct protection of agricultural exportables.

Compared with direct interventions, the indirect interventions had an even stronger impact on agricultural producer incentives for all countries—except Egypt and Malaysia for some years (see Table 3.4). On average, the indirect effects on incentives to farmers were two and one-half times larger than those of direct ones, and for most countries the indirect effects exacerbated the negative direct protection, resulting in extremely large total negative total protection. The magnitude of the (total) negative protection or effective taxation was thus quite large. In 1975–79, producer prices of exportables were half the nonintervention prices, and often even less, in Côte d'Ivoire, Egypt, Pakistan, and Sri Lanka; the same was true in 1980–84 in Argentina, Dominican Republic, Ghana, the Philippines, Sri Lanka, Turkey, and Zambia. Cocoa producers in Côte d'Ivoire, for example, in 1975–79 received about one-third the price they would have received at a realistic exchange rate with no direct intervention; in 1980–84, they received only one-half the price.

Additional farm products (exports and importables) were included in a recent extension of the calculations from the same project. According to these results, in 1960–84 the nominal rates of protection for agricultural exports at the official exchange rate were negative, on average −12 percent, whereas for farm importables they were positive, on average 13 percent.[13] Several countries, including Ghana and Côte d'Ivoire, had an even greater negative nominal rate of protection for agricultural exports—more than −25 percent. The combined effect of direct and indirect price interventions on the principal export products of the eighteen countries during 1960–84 was huge. On average, the implicit export tax equivalent was 34 percent, whereas for agricultural importables the tax was 10 percent. During the 1960–84 period there was an average implicit tax of 30 percent for all farm products.

[13]See synthesis volume no. 4 of this project by Schiff and Valdés (forthcoming).

Two significant results came out of this study. First, there is a marked contrast between the direct policies adopted toward export crops and those directed to import-competing food products; governments tax the production of exportables and protect the production of food. Second, the indirect rate of taxation resulting from economywide interventions is very high; this indirect effect dominates the direct effect (resulting from sectoral policies), whether the direct effect is negative or positive.

An Outward-Oriented Trade Strategy: The New Export Pessimism

The agricultural performance of many developing countries, particularly those in Sub-Saharan Africa, has been disappointing for many years, and it is generally agreed that domestic policies must change if economic performance is to improve. In many LDCs, these agricultural reforms are part of a package of policy reforms usually referred to as a structural adjustment program. The recommended package consists of a mix of demand-side and supply-side policies and specific measures, which include price and nonprice policies, to improve a country's international competitiveness. An especially important role is attached to an outward-oriented trade strategy that would expand agricultural exports.

Among economists and policy makers, however, there is considerable controversy about the appropriateness of such an approach. Concerns about constraints in external demand continue, especially in the slow-growth environment of the world economy in the 1980s. Domestic food security is seen to be threatened if there is a trade-off between expanded agricultural exports and the availability of domestic food supplies. At the local level, fears are also expressed about the nutritional consequences for farm households of switching from food crop to export crop production. And if domestic food prices rise, what will be the impact on the poor?

External Demand Constraints

Economists do not agree whether a strategy of agricultural export-led growth is viable if it is pursued by a large number of countries simultaneously. Two alternative but mutually exclusive hypotheses have been raised. One side argues that external demand constraints could impede growth in export revenues if a jointly promoted export-led strategy resulted in a fall in world prices for the export commodities. Others argue that rigidities in domestic supply responses could preclude significant export growth. Of course, if domestic supply cannot expand under these policies, then any potential external demand constraint would not be binding. It is maintained here that the external demand constraint weighs heavily as an underlying concern in advocating an agricultural export-led strategy.

This external demand constraint has at least two distinct dimensions, and both could result in a decline in world prices. First, traditional tropical products do not compete with domestic production in developed country markets, and consequently they usually face relatively low levels of protection in those markets unless a country exports a commodity in semiprocessed form instead of the raw material. This applies particularly to some exports from Sub-Saharan Africa such as cocoa. Second, exports from temperate and subtropical areas such as South America, some of which—beef, sugar, and some horticultural products, for example—do face high trade barriers in developed country markets even when exported as raw materials.

In the case of Sub-Saharan Africa, the region for which this concern is most acute, the external demand constraint has probably been overestimated as a cause of the failure of structural adjustment policies.[14] Countries in this region have experienced a loss in market share for most commodities over time. A policy that would stop the decline in exports, or at least maintain current market shares, could be termed moderately successful, and such a policy would be likely to have only a marginal effect on world prices. Of course, exports from this region could be further stimulated if industrialized countries such as those in the European Community were to open up their markets to processed agricultural imports. Removing the present tariff escalations could help countries in Sub-Saharan Africa set up export-oriented, agricultural processing industries to produce exports with higher unit value.

The main problems arise for commodities for which Sub-Saharan Africa has a rather large share in the world market—coffee and cocoa, the region's two chief agricultural export products. Significant expansion of coffee and cocoa exports would reduce world prices and perhaps reduce marginal export revenues in those countries. Other products exported from the region, such as palm kernel,

---

[14]This conclusion stems from the analysis by Koester, Schafer, and Valdés (1989) of tropical products exported from Sub-Saharan Africa. This section draws on that study.

groundnut oil, groundnuts, and sisal, also have a large share in world markets, but they represent a fairly small share of the region's export revenues; hence, their lower export price should cause fewer adjustment problems than would lower prices for coffee and cocoa.

Excessive emphasis on two or three export products could be misleading. Countries that have followed open trade regimes have usually succeeded—as did Brazil, Chile, Thailand, Turkey, and others—in diversifying their agricultural exports significantly over time. The terms-of-trade effect on export revenues from traditional export products should induce a shift of some production toward other exportables. Increased investment in research and extension would have a high payoff in reducing production costs and improving the competitiveness of Sub-Saharan Africa in tropical products vis-à-vis Brazil, Malaysia, and others. The future competitiveness of Sub-Saharan African countries among competing exporters of tropical products is very difficult to evaluate; prospects for expanding exports of cocoa and coffee in the short run do not look favorable (Koester, Schafer, and Valdés, 1989).

Countries in Sub-Saharan Africa might face less severe external demand constraints if adjustment policies were simultaneously implemented and were to stimulate intra-LDC trade in agricultural products. In the long run, intra-LDC trade could increase because adjustment policies should reduce trade barriers and are supposed to spur growth in incomes, providing an opportunity for more diversified exports and less dependence on the import demand of developed countries. A comparison of agricultural production and exports among countries of Sub-Saharan Africa reveals that their production and consumption patterns differ considerably (Koester, Schafer, and Valdés, 1989). In 1981–84, the observed intra-area trade was only 9 percent of agricultural exports; however, trade overlap indicators for this time period show that on average these countries spent 23 percent of their export revenues on imports that are at least statistically classified as the same products exported by countries in the region. Even under the assumption of imperfect substitution between these two sources, the potential trade within Sub-Saharan Africa was therefore probably not less than twice its actual 1981–84 level.

To expand trade among African countries is, of course, a very complex task, one that takes years. It requires modifications in the exchange-rate system, simultaneous liberalization efforts across African countries, including removal of quantitative restrictions and licenses, improvement in roads and transport facilities, and other adjustments. Perhaps the most difficult aspect of designing and implementing a sustainable policy reform is to achieve a cohesive package that generates credibility about the policies over the long term; such credibility is critical for stimulating private investment and the necessary efforts to search for new market opportunities.

Cash Crops for Export: Impact on Incomes and Nutrition

A commonly held perception is that small farmers in several low-income developing countries who produce crops for export, compared with subsistence farmers who produce basic food crops for home consumption, expose themselves to excessive risks in terms of food security and nutrition. A less paternalistic view would suggest that farmers are very aware of their food risks and would choose to balance the expansion of cash crop production with their food needs, a view supported by recent research in Africa, Asia, and Latin America.[15] Small farmers who shifted into sugar cane, maize, vegetables, and rice for export, maintained or slightly improved the nutrition levels of their families while raising their incomes (von Braun, 1988). According to this research, farmers continue to produce subsistence crops as insurance against market and production risks. This concern for food security creates an additional demand by farmers for improved technology for production of staple foods if production of cash crops for export is to be expanded.

Another very important finding of this research is that a major by-product of the switch to an export-oriented agriculture was an increased demand for hired labor, thus creating jobs. The researchers concluded that farmers who grew cash crops benefit in the long run; in fact, introducing export-oriented cash crops into agriculture is probably a necessary part of the long-term development process.

Risks to National Food Security

Government officials in many developing countries perceive a dilemma between an agricultural export-oriented strategy and the risk of food shortages. This dilemma is vividly illustrated by the experience of Kenya, which controlled the export of pulses, cereals, and

[15]Cross-country studies, which were coordinated by Joachim von Braun of the International Food Policy Research Institute, were based on farm and household surveys conducted over a three-year period in several areas, including the western highlands of Guatemala.

meat during the early 1980s. "The political and economic risks of food exports are reflected in the unwillingness of even senior civil servants [in Kenya] to authorize exports of specific commodities within their jurisdiction. Permission to export has to go to the very highest level of government for approval. . . . The pressures on the time of senior politicians make for slow decisions, whereas the availability of the crop itself, and the international prices offered, may change rapidly"(Schluter, 1984, p. 102). When such uncertainty about being able to export is widespread among farmers and traders, specialization for the export market cannot take place. More broadly, a concern for food security makes policy makers apprehensive about exposing a country to the risks of a trade-oriented agriculture.

Trading in International Markets.   One of the most serious objections to liberalized agricultural trade is that it increases the dependence of a country on foreign supplies for part of its food needs. Exporting some nonfood agricultural products and using the proceeds to import food is thought to be risky, because grain supplies in the world market are highly unreliable. To evaluate this risk, one must sort out several factors. Although commodity prices in the world market are highly unstable, supplies are always available for the needs of small- and medium-sized countries—with the exception noted below. The reliability of concessional as opposed to commercial supplies also differs considerably. Concessional supplies are certainly unreliable, and thus national food policies relying on food aid are highly risky. But it does not follow that each country should therefore become self-sufficient in the basic food staple. To defend that position, one would have to show that commercial supplies are also unreliable.

A country relying on commercial supplies assumes a risk if its basic food staple is not widely traded in the world market. There has never been a time, for example, when wheat could not be obtained at some price, even though the price may have been too high for some poor countries. For rice and white maize, however, there have been periods, such as 1973–74, when supplies were unavailable at any price. Markets for both rice and white maize are "thin" and therefore subject to sudden interruptions and shifts in trade channels. It may be to the advantage of countries whose basic food staple is rice or white maize to set the domestic price of the basic foodstuff somewhat higher than the border price to reduce imports, thus reducing the country's vulnerability to shortages in the world market. In such cases, the design of a food policy would be highly specific to each country.

Certain government policies may be required to reduce the risks of trading in international markets, thus encouraging their use by private traders. It is commonly presumed that a more open economy is more specialized and therefore riskier, but the evidence that this is true is weak (Valdés and Siamwalla, 1988). On the contrary, when a more open regime is introduced, the export composition becomes more diversified. Examples abound, such as Thailand in the 1960s and Chile, Turkey, and Greece in the late 1970s. Nonetheless, international trade involves risks, and private producers are, in fact, risk-averse. Private agents will therefore tend to avoid making full use of their comparative advantage, reducing their exposure to trade and consequently causing some loss in efficiency to the economy. In the absence of private institutions to share risk, the main corrective action required from the government is to mediate the sharing of risks. The government could intervene to stabilize domestic prices within a band by implementing variable trade taxes and a subsidy scheme (Siamwalla, 1986).

Maintaining Incentives for Agricultural Production and Exports. Trade liberalization is likely to alter sharply the terms of trade between commodities produced for domestic food consumption and those produced for export. Because of the often significant overvaluation of the domestic currency, especially in the countries of West Africa, and the prevalence of export taxes and licenses, liberalization will stimulate the production of exportables, possibly at the expense of local food supplies. To maintain adequate domestic food consumption, the country will have to import food, and this could undermine its food security.

Particularly for West Africa, an argument for food security based on a policy goal of agricultural import substitution favors a selective closure of the food economy, thus implicitly taxing the production of agricultural exports. Behind this argument made by Aboyade (1990) is a presumed "import incapacity" of the countries, which does not allow sustainable imports of food over a long period of time. This prescription to limit agricultural trade stems from various conditions. First, the external demand prospects for West Africa's agricultural exports are not seen to be favorable. Second, production of cash crops and production of food crops are regarded as trade-offs rather than complements. The potential benefits of a policy that encouraged cash crop production for export would be dissipated by the negative impact on domestic food production and nutritional status. Third, there is a lack of congruency between the food items produced and those consumed, resulting in growing food deficits; dietary habits and consumption have rapidly shifted away from traditional staple foods toward importables.

Such a prescription to impose trade barriers implies direct protection to food crops. Most developing countries do directly protect products that are importables and tax agricultural exports.[16] But in parts of Africa, the major staples are nontradables; their production is not subsidized and their consumption is (implicitly) discouraged. A direct subsidy to production of nontradables requires fiscal expenditures, in contrast to the protection of importables and taxation of exportables, which generate government revenues. But economywide policies generally dominate the direct effects on agricultural incentives, whether these effects are positive or negative.[17] The net result of the combined effect of sectoral and economywide policies on the domestic market of agricultural tradables is that, relative to nontradables, their production has been taxed and their consumption has been subsidized, mainly because of the overvalued domestic currency.

Hence, the import incapacity predicted by Aboyade (1990) and others for West Africa is perhaps largely endogenous to the policy process. A country's import capacity is directly related to the total effects of price interventions on the foreign exchange rate and, through this, on the net value of agricultural trade. The overall effects of price changes that result from policy interventions depend on both consumption and output effects, as well as any changes in use of tradable inputs. When Krueger, Schiff, and Valdés (1988) studied these effects for the 1960 to 1985 period by simulating the removal of all direct and indirect price interventions, they concluded that the actual policies pursued in most of the eighteen countries in their sample caused large foreign exchange losses in net agricultural trade. These losses are due to the effect of direct and indirect taxation on the production of exportables, the indirect taxation of food production (vis-à-vis nonagriculture), and the implicit consumption subsidy in tradable foods relative to the traditional staples, which in most of Africa are usually nontradables. In a further example, for Ghana,

---

[16]Most of the eighteen countries analyzed in Krueger, Schiff, and Valdés (1988) have turned the within-agriculture terms of trade (at the official exchange rate) in favor of import-competing food production and against exportables.
[17]For the empirical evidence, see Krueger, Schiff, and Valdés (1988).

Stryker (1990) estimated that the average annual long-run total foreign exchange loss in agricultural trade from price interventions for 1963–84 was about 80 percent of Ghana's total exports. In the 1960s, these losses averaged 7.6 percent of total export revenues. In the 1970s, they averaged 19 percent; in 1980–84, 320 percent.[18] Foreign exchange revenues would thus have been over four times larger in 1980–84 if all price interventions had been removed starting in the early 1960s. Any lack of import capacity thus stems directly from the policies pursued by the government, not from fundamental structural impediments in either the domestic or the world economy.

In summary, though there is a good case for selective nominal protection to production of thin-market tradables, such as white maize in East Africa and rice in Asia, a more efficient remedy for the predicted import incapacity of the food sector is to correct the exchange rate. Such a policy would eliminate, on the consumption side, the implicit subsidy on importables and, on the production side, the taxation of agricultural tradables.

The Output Response of Agriculture

The high negative rates of protection for agricultural exportables, as presented earlier, indicate the scope the government has for reversing course and offering price incentives. Implicit in much of the literature on infant industry, however, is the assumption that agriculture is destined for a static role technologically, whereas industry is dynamic. Although crop output of individual farmers responds to price movements, the aggregate supply of agricultural products from the sector as a whole is thought by some to be unresponsive to incentives—the so-called aggregate supply inelasticity of agriculture in developing countries. If that assumption is correct, the social cost of using agriculture as a tax base for economic development is low.[19] That assumption is challenged here. One expects the aggregate supply response to price movements to be lower than that of individual crop output, because it costs more to switch the resources required for aggregate supply response between sectors than to switch resources between crops. But favoring industry on the grounds that the

[18]These are net effects considering consumption, production, and use of tradable inputs and adjusting for world price effects on cocoa.

[19]The term *agricultural taxation* is used in this chapter to refer not to land or income taxes but to trade taxes and quantitative restrictions, or, more generally, the terms of trade of agriculture vis-à-vis the rest of the economy.

price responsiveness of aggregate agricultural output is low is a mistake. The relationship between the aggregate supply response of agriculture and incentives is complex and not yet fully understood. Empirical work on the long-run responsiveness of agriculture has only recently begun to challenge the narrow approach taken in the past in so much of the literature on supply response.

This complexity arises from the difficulty of incorporating intersectoral resource flows and productivity changes that could occur as a result of changes in the domestic terms of trade. The long-run supply response is the sum of the short-run response (in which land, labor, and capital are fixed) and the effect of the price change on output via its impact on the intersector reallocation of "fixed" factors. The key distinction between short-term and long-term output response to price is thus explained by intersectoral flows of labor and capital and by the relationship between incentives and new technology. In the long run, the rate of labor migration depends on intersectoral differences in income. Influenced by real wage differentials between these sectors, labor will move between agriculture and nonagriculture. A similar pattern holds for capital. Given a total level of investment for the whole economy, the sectoral allocation will be influenced by prospective returns to capital in each sector. The flows of both labor and capital from agriculture to nonagricultural sectors should therefore accelerate when prospective returns to these factors in the nonagricultural sectors increase relative to those in agriculture. Such increases will occur when the real exchange rate appreciates—that is, the price of tradables declines relative to that of nontradables.

The analysis of the relationship between prices and the generation and adoption of new technology in agriculture indicates that technical changes in agriculture (tubewells, fertilizer, electricity, equipment) usually require an increase in capital stock, and the adoption of such techniques therefore depends on the rate of capital accumulation and incentives.

Recent work on aggregate agricultural supply response—a measure of supply response using a fuller specification of rural-urban linkages in the labor and capital markets—is beginning to challenge the pessimistic view of the supply response of the agricultural sector. Some of the best technical work in this field has been done on South American countries, specifically the studies by Cavallo and Mundlak (1982) on Argentina, and by Mundlak and Coeymans (forthcoming) on Chile. Unfortunately, empirical analysis with this economywide view of returns is still not available for Sub-Saharan Africa or Asia. It has often been presumed, particularly for the countries of Sub-Saharan Africa, that agricultural net exports would not respond much to price incentives because of rigidities and inflexibilities in the economic structure.

The results of an important recent empirical study by Balassa contradict such a presumption.[20] According to this study, agricultural exports and net exports are highly responsive to the real exchange rate through its effects on relative agricultural prices. Macroeconomic and trade policies that lead to a fall in the real exchange rate thus have a negative effect on net agricultural exports and, consequently, on foreign exchange earnings. Balassa finds that agricultural exports are more responsive to price incentives in Sub-Saharan Africa than in developing countries in general. How aggregate agricultural supply (not only exports) responds to changes in the real exchange rate in this region of Africa is an empirical question, however, which has not been addressed satisfactorily.[21]

Conclusion

After decades of experience in developing countries with inward-oriented development strategies, which grossly undervalued the potential contribution of exports to development, governments are reassessing the positive contribution of exports, including agricultural exports, to economic growth. The industrial growth strategy through import substitution has not fulfilled the expectations of its

[20]Balassa (1988) examined the relationship between changes in the real exchange rate and changes in the agricultural export-to-output ratio for fifty-three developing countries, including sixteen Sub-Saharan African countries (the analysis held constant the level of foreign incomes, i.e., the combined GNPs of the developed countries as a control for changes in external demand). For the 1965–82 period, he found an elasticity of the agricultural export-output ratio with respect to the real exchange rate equal to 0.68 for the entire group of countries and equal to 1.35 for the Sub-Saharan African countries analyzed (both significant at the 1 percent level). A similar relationship was examined when the dependent variable was net exports minus imports. For 1965–82, the elasticity of the net export-output ratio with respect to the real exchange rate was 4.96 for the developing countries analyzed and 11.5 for the sixteen Sub-Saharan African countries.

[21]Most of the literature is dominated by studies using a single equation time series approach, which fails to capture the underlying migration and investment processes. See Bond (1983), Herdt (1970), Peterson (1979), Reca (1980), Krishna (1982), and Chhibber (1982). The supply response depends on wages and rates of return in agriculture, which are not captured in these studies. It also depends on wages and rates of return in other sectors, since they affect the supply of inputs—these are not captured in cross-country analysis either. Also left out of the empirical estimates are expectations and price variability, which could be very influential on the final outcome.

advocates in most of the developing countries in which it has been followed. Faster growth of exports is, of course, crucial for highly indebted developing countries. The countries now experiencing rapid economic growth are those that have achieved vigorous growth in exports.

In addition to the contribution of exports to foreign exchange earnings, some governments, particularly in low-income countries, have recognized other benefits and have shown increasing interest in promoting agricultural exports as part of an agriculture-based development strategy. Agricultural progress is increasingly regarded as a strategic element for the development process in low-income economies. Such an export-led strategy has enormous potential to generate employment and thus help alleviate poverty. There are, however, several current concerns about the adoption of an outward-oriented strategy for agriculture. Dependence on agricultural exports, particularly by food-deficit countries, is seen as undesirable. Although these concerns have to be assessed in the particular context of an individual country, in most cases pessimism about export-oriented strategies is unwarranted.

Policy reform is urgently needed. In the past, agricultural exports have been heavily taxed in most developing countries—on average at a rate of approximately 35 to 40 percent—as a result of the combined effect of sectoral and economywide policies. A need for government revenue was clearly an influential factor underlying the direct taxation of exports in many developing countries.[22] But the most pervasive form of taxation came from misalignment of the real exchange rate, which did not significantly generate government revenue. Indirect taxation of agricultural exports has been substantial, and the removal of these indirect taxes requires reform of economywide policies. For there to be growth in agriculture, the structure of incentives has to change. Recent evidence, including that for Sub-Saharan Africa, points to the strong response of agricultural exports to incentives.

An outward orientation is not a matter of trade policy alone. Sustained growth of a broad base of agricultural exports is dependent on an organizational framework and "trade infrastructure," which in some developing countries are not in place. Export growth is

---

[22]The lowering of tariff rates on industrial products can have a negative revenue effect, at least in the short run. To the extent that quantitative import restrictions are replaced by tariffs, even at relatively low levels, however, the revenue loss from import liberalization is reduced.

dependent on an efficient service sector, including banking and communications, and on a regulatory framework for trade and improved physical infrastructure. It takes several years to put these elements in place. Until then, developing countries cannot hope to benefit from adopting a more outward orientation for agriculture. For economic and social reasons alike, it is important that developing countries without an effective economic and institutional framework move ahead to initiate the necessary reforms. There is much to be lost and very little to be gained by waiting until the last moment when a case for reform can be made.

Acknowledgments

I gratefully acknowledge the comments on an earlier draft by Romeo Bautista, Chris Delgado, and Peter Timmer.

References

Aboyade, Ojetunji. 1990. "Perspectives on African Food Policy and Agricultural Development in the 1990s." In Neil G. Kotler, ed. *Sharing Innovation: Global Perspectives on Food, Agriculture, and Rural Development*. Washington, D.C., and London: Smithsonian Institution Press.

Balassa, Bela. 1984. *Adjustment to External Shocks in Developing Countries*. World Bank Staff Working Paper 472. Washington, D.C.: World Bank.

——. 1988. *Incentive Policies and Agricultural Performance in Sub-Saharan Africa*. World Bank Staff Working Paper 77. Washington, D.C.: World Bank

Bautista, Romeo M., and Alberto Valdés, eds. Forthcoming. *Trade and Macroeconomic Policies in Developing Countries: Impact on Agriculture*. Baltimore: Johns Hopkins University Press.

Bond, Marian E. 1983. "Agricultural Responses to Prices in Sub-Saharan African Countries." *IMF Staff Papers* (December): 703–26. Washington, D.C.: International Monetary Fund.

Cairncross, A. K. 1962. *Factors in Economic Development*. London: George Allen and Unwin.

Cavallo, Domingo, and Yair Mundlak. 1982. *Agriculture and Economic Growth in an Open Economy: The Case of Argentina*. IFPRI Research Report 36. Washington, D.C.: International Food Policy Research Institute.

Chhibber, Ajay. 1982. *Dynamics of Price and Non-Price Response of Supply in Agriculture*. Stanford: Stanford University Press.

Feder, Gershon. 1982. "On Exports and Economic Growth." *Journal of Development Economics*. 12, nos. 1–2:59–73.

Herdt, Robert W. 1970. "A Disaggregate Approach to Aggregate Supply." *American Journal of Agricultural Economics* 52, no. 4:512–20.

Hirschman, Albert O. 1958. *The Strategy of Economic Development*. New Haven: Yale University Press.

Islam, N., and A. Subramanian. 1989. "Agricultural Exports of Developing Countries: Estimates of Income and Price Elasticities of Demand and Supply." *Journal of Agricultural Economics* 40, no. 2:221–31.

Koester, Ulrich, Hartwig Schafer, and Alberto Valdés. 1989. "External Demand Constraints for Agricultural Exports: An Impediment to Structural Adjustment Policies in Sub-Saharan African Countries?" *Food Policy* 14, no. 3:274–83.

Krishna, Raj. *1982* "Some Aspects of Agricultural Growth, Price Policy and Equity in Developing Countries." *Food Research Institute Studies* 18, no. 3: 219–60.

Krueger, Anne O. 1978. "Alternative Trade Strategies and Employment in LDCs." *American Economic Review* 68, no. 2:270–74.

———. 1980. "Trade Policy as an Input to Development." *American Economic Review* 70, no. 2:288–92.

———, ed. 1988. *Development with Trade: LDCs and the International Economy*. San Francisco: ICS Press.

Krueger, Anne O., Hal B. Lary, Terry Monson, and Narongchai Akrasanee, eds. 1981. *Trade and Employment in Developing Countries*, vol. 1. Chicago: University of Chicago Press for National Bureau of Economic Research.

Krueger, Anne O., Maurice Schiff, and Alberto Valdés. 1988. "Agricultural Incentives in Developing Countries: Measuring the Effect of Sectoral and Economywide Policies." *World Bank Economic Review* 2, no. 3:255–72.

Lewis, Stephen R. 1989. "Primary Exporting Countries." In Hollis Chenery and T. N. Srinivasan, eds., *Handbook of Development Economics*, vol. 2, pp. 1541–1600. Amsterdam: North-Holland.

Little, Ian M. D., Tibor Scitovsky, and Maurice Scott. 1970. *Industry and Trade in Some Developing Countries: A Comparative Study*. London: Oxford University Press for OECD.

MacBean, Alasdair. 1966. *Export Instability and Economic Development*. Cambridge: Harvard University Press.

———. 1989. "Agricultural Exports of Developing Countries: Market Conditions and National Policies." In Nurul Islam, ed., *The Balance between Industry and Agriculture in Economic Development*, pp. 101–28. London: Macmillan.

Meier, Gerald M. 1975. "External Trade and Internal Development." In Peter Duignan and Lewis H. Gann, eds., *The Economics of Colonialism*. London: Cambridge University Press.

Mellor, John W. 1976. *The New Economics of Growth: A Strategy for India and the Developing World.* Ithaca: Cornell University Press.

Mundlak, Yair, and Juan Eduardo Coeymans. Forthcoming. "Agricultural and Sectoral Growth: Chile, 1962–1982." In Romeo M. Bautista and Alberto Valdés, eds., *Trade and Macroeconomic Policies in Developing Countries: Impact on Agriculture.* Baltimore: Johns Hopkins University Press.

Myint, Hla. 1955. "The Gains from International Trade and the Backward Countries." *Review of Economic Studies* 22, no. 2:129–42.

———. 1958. "The Classical Theory of International Trade and the Underdeveloped Countries." *Economic Journal* 68 (June): 317–31.

Nurkse, Ragnar. 1961. "Trade Theory and Development Economics." In Howard S. Ellis, ed., *Economic Development for Latin America.* New York: St. Martin's Press.

Peterson, Willis L. 1979. "International Farm Prices and the Social Cost of Cheap Food Policies." *American Journal of Agricultural Economics* 61, no. 1:12–21.

Reca, Lucio. 1980. *Argentina: Country Case Study of Agricultural Prices and Subsidies.* World Bank Staff Working Paper 386. Washington, D.C.: World Bank.

Schiff, Maurice, and Alberto Valdés. Forthcoming. *The Economics of Agricultural Price Interventions in Developing Countries.* Baltimore: Johns Hopkins University Press.

Schluter, Michael. 1984. *Constraints on Kenya's Food and Beverage Exports.* IFPRI Research Report 44. Washington, D.C.: International Food Policy Research Institute.

Siamwalla, Ammar. 1986. "Approaches to Price Insurance for Farmers." In Peter Hazell, Carlos Pomareda, and Alberto Valdés, eds. *Crop Insurance for Agricultural Development: Issues and Experience.* Baltimore: Johns Hopkins University Press for IFPRI.

Stryker, J. Dirck. 1990. *Trade, Exchange Rate, and Agricultural Pricing in Ghana.* World Bank Comparative Studies: The Political Economy of Agricultural Pricing Policy. Washington, D.C.: World Bank.

Timmer, C. Peter. 1988. "The Agricultural Transformation." In Hollis Chenery and T. N. Srinivasan, eds., *Handbook of Development Economics,* vol. 1, pp. 275–333. Amsterdam: North-Holland.

———. 1989. "Food Price Policy: The Rationale for Government Intervention," *Food Policy* (February): 17–27.

Timmer, C. Peter, Walter P. Falcon, and Scott R. Pearson. 1983. *Food Policy Analysis.* Baltimore: Johns Hopkins University Press for the World Bank.

Valdés, Alberto, and Barbara Huddleston. 1977. *Potential of Agricultural Exports to Finance Increased Food Imports in Selected Developing Countries.*

IFPRI Occasional Papers 2. Washington, D.C.: International Food Policy Research Institute.

Valdés, Alberto, and Ammar Siamwalla. 1981. "Introduction." In Alberto Valdés, ed., *Food Security for Developing Countries,* pp. 1–24. Boulder, Colo.: Westview Press.

——. 1988. "Foreign Trade Regime, Exchange Rate Policy, and the Structure of Incentives." In John W. Mellor and Raisuddin Ahmed, eds., *Agricultural Pricing Policies for Developing Countries.* Baltimore: Johns Hopkins University Press.

von Braun, Joachim, with Eileen Kennedy and Howarth Bouis. 1988. "Commercialization of Smallholder Agriculture: A Comparative Analysis of the Effects on Household Level Food Security and Nutrition and Implications for Policy." Washington, D.C.: International Food Policy Research Institute, October 1988. Typescript.

4

Observations on Export-Led Growth as a Development Strategy

James P. Houck

A striking theme in much recent literature on economic development in the Third World is that export-promoting policies are now viewed as the touchstone of growth (Krueger, 1988; Valdés and Siamwalla, 1988). Chapter 3 by Valdés clearly summarizes this position and emphasizes the role of agriculture and agricultural trade. The core of the chapter is an empiricial analysis of the performance and structure of agricultural exports in recent years. Valdés also elaborates on the evolution away from the negative ideas about foreign trade, the anti-agricultural export bias, that dominated previous development thinking and policy. Elsewhere in the literature such policies may be termed *import-substitution strategies*. To be considered here is the question of export-led growth from a general perspective. What is export-led growth as an economic policy, and how is it pursued?

Export-led growth is a term that is most often applied to macroeconomic development policy in nations having a large existing export sector or substantial potential for production of exports. In the Third World this potential is most often, but not exclusively, agricultural. The fundamental idea is to set forces in motion that will expand the excess supply of exportable goods, thereby leading to a rapid expansion of output, employment, and earnings in the export sectors. Enhanced activity in these sectors then will spill over into the input and capital markets linked to the tradable goods and then to the economy generally (Krueger, 1980).

A central idea in such a development strategy is to employ the purchasing power of the entire trading world as a major engine of growth rather than rely on the slower accretion of domestic incomes. In addition, vigorous expansion of exports will generally lead to growth in foreign exchange earnings, which then can be deployed both to purchase capital goods and services from abroad and to service international debt. Some proponents of export-led growth also see expanded foreign trade as a major source of new ideas, innovation, and freshness for tradition-bound societies.

Export-Promoting Policies

As a specific policy undertaken broadly by a government, export-led growth can be pursued by various means, some of them efficient from the point of view of most economists and some not. On the one hand, it is unlikely that most nations individually embarking on or contemplating an export-led growth strategy will be able to affect international prices or other market factors directly by increasing or decreasing the volume of products supplied to world markets. On the other hand, export sales and production can be enhanced across the board by three broad classes of deliberate interventions. The first is to dismantle or reduce systematically the trade barriers affecting exports themselves or imported inputs for the export sector. Second, a country can alter the international terms of trade facing buyers and sellers either by means of adjustments in the exchange rate carried out by the nation's central bank or by centrally funded export subsidies. Third, policy makers can make changes in both general and specific domestic economic policies that are designed to promote exportable output and cut production costs in those sectors. Let us consider each of these interventions separately.

Dismantling Trade Barriers

Export growth can be fostered by reducing export taxes, quantitative restrictions on trade volume, export licensing fees and red tape, restrictions on export shipping, and similar impediments. In addition, exports can be expanded by reducing tariffs, quotas, and other import restrictions levied on products that are inputs to or raw materials for export production. Most disinterested economists, including Valdés, applaud without hesitation such policy changes taken in the name of export-led growth. These moves tend to increase economic efficiency and allow patterns of underlying comparative advantage to assert themselves.

Exchange Rates and Export Subsidies

To promote the growth of exports by exchange-rate interventions, policy makers must devalue the nation's currency with respect to those of potential importers and competing exporters. It is also possible that exchange rates for specific export sectors and products can be altered as part of a multiple exchange-rate regime. Because the currencies of many developing countries have been traditionally and chronically overvalued, export-promoting devaluations can intensify pressures for domestic inflation by suddenly pushing up prices of imports across the board and by increasing domestic prices of goods that are also exported.

Direct export subsidies have effects on trade and prices that are similar to currency devaluation. Prices of export goods on the domestic market tend to rise, whereas prices paid by foreigners hold steady or fall. Trade volume and foreign exchange earnings tend to increase. Real resources of the central government, however, must be allocated away from other uses and put into payments of export subsidies. Similarly, subsidized export production pulls resources out of other parts of the domestic economy, which entails efficiency losses that are well known.

As a general rule, economists favor the elimination of deliberate currency overvaluation but are less enthusiastic about artifical devaluation or multiple exchange-rate regimes as a spur to export activity. On balance, economists also find export subsidies to be onerous. They promote the inefficient internal allocation of resources and punish domestic consumers of export goods and their close substitutes with higher prices.

Internal Policies

One could list dozens of possible internal policy adjustments that could be applied to favor the export-producing industries and sectors. Three main ones offer sufficient illustration: wage and labor policy, tax policy, and policy to alter the cost of purchased inputs.

Hired workers in industries targeted for export expansion in most Third World countries can be expected to experience downward pressure on their wage rates or strong resistance to increases. Wage policy generally will attempt to keep labor costs down and output prices competitive on international markets. In autocratic and centrally planned economies, policy makers will try to hold low-wage labor in the export sectors, discouraging its natural out-migration.

Tax policy obviously can favor export sectors with favorable treatment in many dimensions to encourage both current profits and long-run capital investments. Input subsidies, preferential access to factor supplies and credit, cost rebates, and domestic consumption (excise) taxes are policy measures that tip the scales of market operation deliberately in favor of expansion in production destined for export. Overall, owners and managers of productive resources in export sectors can expect to experience favorable returns as export-led growth is pursued.

Disinterested economists voice differing views on the merit of these internal measures because of differing social philosophies as well as specific assessments of countries, products, current conditions, past experience, and international markets. It is widely agreed, however, that the heavy tax burden carried by agriculture, one way or another, in numerous less-developed nations is a considerable barrier to export expansion. Valdés is clear about this point.

Agriculture's Role

For the sake of discussion, let us assume that a heavily agricultural, less-developed nation follows the advice implicit in Chapter 3 and, using the policy tools available, embarks on an extensive and serious program of export-led growth. What role will its agriculture play in such an effort, and what results can be expected internally? To probe this question, even briefly, we need to set out some ideas about the structure of this economy and its agriculture. Let us assume the following structure, which is a recognizable prototype of many Third World nations:

A large proportion of the population is employed in agriculture; many of them are landless workers. Unemployment exists among these farm laborers.

The agricultural sector contains two major parts, a sector producing food grains and feed grains and a plantation enclave producing raw material cash crops. Most other foods are produced domestically, but some imports of food and livestock feed are usually needed.

Most of the output of plantations and some of the output of grain is exported.

A traditional oligarchy owns large tracts of land in a plantation enclave. Other farmland is more widely held in small units.

Mechanical and chemical inputs for farming are mostly imported and subject to tariffs at the border.

National per capita GNP is sufficiently low that a large portion of disposable income must be spent for food.

Now let us suppose that the export-led growth policy has the following features: across-the-board devaluation of the currency, with additional devaluation applied to exports of cash crops under a multiple exchange-rate regime; removal of the traditional export taxes on grain and raw materials; lower wage rates for rural landless workers and internal migration controls to prevent population growth in the cities; removal of import tariffs on farm machinery, fertilizers, and pesticides; and internal cuts in the annual land tax levied on all landowners. How would this nation's agriculture fare under such a program? Moreover, what major implication would such a program have for the rest of the economy? To focus this illustration further, let us presume that world markets for grains and for the nation's other cash crops remain stable and generally favorable. For this hypothetical country, this policy setting would wipe out both direct and indirect negative protection for agriculture of the sort illustrated in Chapter 3 (Table 3.4).

One cannot be very precise or emphatic about the aggregate and differential effects to be achieved, but a few qualitative generalizations can be advanced. In the first place, all of these policies individually tend to expand production and total exports of grain and cash crops. If the nation is a small trader in world markets, or if its export demand is price elastic in foreign currency, then foreign exchange earnings would rise. The special devaluation applied to plantation cash crops would provide a positive additional stimulus to that sector.

All across agriculture, prices and gross earnings of producers, measured in domestic currency, would tend to rise. Of course, prices of grains and cash crops quoted in foreign currencies would remain steady or fall. Prices of imported inputs might rise or fall in terms of the domestic currency depending upon the extent to which devaluation and decreases in tariffs offset each other. In any case, farm labor costs would hold steady or fall.

With prices of outputs on the rise, costs of inputs steady or falling, and taxes decreasing, net income to landowners and farm operators would increase. Their resulting prosperity could be expected to spill over into other sectors of the economy as their spending on production inputs and consumption goods expands. Increases in income to an already wealthy elite, however, also might drain into unproductive purchases of imports or overseas bank accounts and be lost for development. As posited here, the employed landless workers in agriculture would seem to benefit little from such export-led growth, but new low-wage jobs in the farm sector probably would become available for previously unemployed or underemployed workers.

It seems likely that food prices would increase throughout the economy at large, especially if tariffs on imported foods were left in place. Food grains, livestock products, and other farm products likely would be affected. Much in this policy package points toward inflation, at least for the short term. Increases in food prices over time might abate if cheaper costs for purchased inputs and labor allowed for supply expansion and lower prices of exports. In addition, if the entire program is successful in creating income spillovers into the economy at large, the proportion of per capita disposable income spent on food might, on average, hold steady or even fall despite higher food prices. Still, general price inflation remains as a lurking consequence of export-led growth. Currency devaluation itself is inflationary throughout an economy, and the differential effect on food prices of this policy package adds to the potential for inflation.

From a purely fiscal point of view, the government would see an immediate drop in its revenue caused by reduced export taxes, import tariffs, and land taxes. The internal politics of such fiscal setbacks are likely to impede aggressive policies for export-led growth in this or any other country. Increased tax revenues from elsewhere in the economy surely would be required in the short run. Over the longer run, replacement revenues might be generated by the expansion of general economic activity stimulated by vigorous export growth.

Conclusion

No doubt, many more conjectures could be spun out of this plausible illustration or variations upon it. Export-led growth, as a sensible choice of development policy, is available to countries with one or more well-integrated tradable goods sectors. Preexisting trade barriers (especially against exports), heavy internal taxes on tradables, and an overvalued currency allow for maneuverability into such a policy.

However, if factor prices are held down by controlling wages for large numbers of people, if general price inflation increases markedly, or if government revenues decline precipitously, then export-led growth would be difficult to sustain. Fears of such effects will and do stand in the way of outward-looking policy reforms. Perhaps the modern experience of trade-oriented countries such as Korea and Taiwan, now in the category of newly industrialized countries, and some others like Thailand and Malaysia, now developing rapidly, will signal the value of a more open trading regime across the rest of the Third World.

References

Krueger, Anne O. 1980. "Trade Policy as an Input to Development." *American Economic Review* 70 (May): 288–92.

———, ed. 1988. *Development with Trade: LDC's and the International Economy.* San Francisco: ICS Press.

Valdés, Alberto, and Ammar Siamwalla. 1988. "Foreign Trade Regime, Exchange Rate Policy, and the Structure of Incentives." In John W. Mellor and Raisudden Ahmed, eds., *Agricultural Price Policy for Developing Countries,* pp. 103–23. Baltimore: Johns Hopkins University Press.

5

Agricultural Employment and Poverty Alleviation in Asia

C. Peter Timmer

Most analysts would agree with the viewpoint expressed by Hayami and Ruttan that inappropriate policies are the major constraint limiting the pace of agricultural development in the developing world. "The basic factor underlying poor performance was neither the meager endowment of natural resources nor the lack of technological potential to increase output from the available resources at a sufficiently rapid pace to meet the growth of demand. The major constraint limiting agricultural development was identified as the policies that impeded rather than induced appropriate technical and institutional innovations. As a result, the gap widened between the potential and the actual productive capacities of LDC agriculture" (1985, p. 416). The extent of that agreement, however, varies for different components of agricultural policy. In some areas there is broad consensus on the role of government involvement—for example, in the provision of such public goods as research and extension and investments in rural infrastructure and irrigation. The debate is at the margin over levels of funding, institutional organization, and balance between the roles played by the private and public sectors. Much of the analysis conducted by Hayami and Ruttan focuses on this debate.

Significant areas of government intervention into agriculture do not reflect such consensus, however. Despite agreement that government policy is important, its efforts to alter the distribution of land

and its tenure arrangements, to stimulate (and control) farmer organizations and cooperatives, to operate marketing boards, and to set domestic agricultural prices, for example, are controversial in both intent and design. Intervention in these areas is motivated by several important objectives, especially to reduce rural poverty, to generate employment, and to enhance food security by stabilizing food prices. None of these objectives, proponents argue, can be achieved by the private sector alone. The reduction of poverty and improved distribution of incomes, the stimulation of employment creation to meet the needs of a rapidly expanding work force (in both urban and rural areas), and the stabilization of the domestic economy in the face of uncertain harvests and volatile world markets thus require an active public role. At a minimum, this role complements that of the private sector. But governments often feel compelled to intervene to demonstrate a commitment to these social objectives that lie at the heart of a government's political legitimacy.

Unfortunately, measures undertaken by the government to create jobs and reduce poverty are often counterproductive. On occasion, government policies for both the macro economy and the agricultural sector are so bad that agricultural output declines, inflation runs out of control, and black markets become the major source of provisions for the nonfarming population. In other circumstances, however, governments have intervened skillfully to stabilize the economic environment and stimulate rapid growth in the agricultural sector. The effect has been to increase demand for labor, pull up wages for unskilled workers, and permit substantially higher levels of caloric intake among the poor, a tangible marker of reduced poverty. Obviously, there is a *potential* role for government interventions to speed the development process in ways that also reduce poverty.

This potential is explored here by analyzing government efforts to stimulate employment in agriculture and examining their impact on poverty alleviation. Most of the evidence presented is from Asia, where most of the success stories with respect to government policy for agriculture can be found. There is plenty of negative experience from the Asian region as well, however, and the comparative experience within Asia is especially helpful in discovering the common elements of good policy. At the same time, restricting the analysis to a single continent controls, although only partially, for the many sharp contrasts in history, culture, and ecological environment that would have to be included if, for example, African and Latin American experiences were compared.

This evaluation extends beyond the agricultural sector into macroeconomic and trade policies. A key lesson of postwar experience, summarized in Chapter 3 by Valdés, has been the importance of links between agriculture and the rest of the economy in stimulating the appropriate technical and institutional innovations that Hayami and Ruttan observe are the basis of successful agricultural development. Consequently, the impact of government policies on employment creation in agriculture cannot be understood in the context of sectoral policies alone. The mechanisms that raise real wages for unskilled workers in rural areas involve crop prices, rural public works, and agricultural technology, of course. But equally critical are the efficiency of links between rural and urban labor markets and the extent to which government policy facilitates labor absorption in manufacturing, urban construction, and the informal service sector. One clear lesson since the 1960s is that countries with the best record of alleviating poverty also have good records in raising rural wages by actively connecting labor markets in rural and urban areas. This "Asian connection" argues that agricultural development is a multisectoral task. Narrowly focused agricultural or rural development policies, even if sound within their own context, are relatively ineffective without complementary policies for trade, finance, industry, and the macro economy.

Efforts to increase real wages through the stimulation of rural employment, if conducted in an environment of reasonable price stability, have a joint impact on food security for the poor. Rising real wages and stable food prices are almost certain to enhance the capacity of poor households to gain access to staple foods on a regular basis. Empirically, this enhanced food security for the poor should be reflected in rising average levels of caloric intake for the entire society. Members of middle- and upper-income households increase their caloric intake only slightly as incomes increase (although the composition of the foods that provide these calories can change substantially); caloric intake of the poor rises if, and only if, the average for the whole society rises. Fortunately, statistics are available to examine the extent to which average caloric intake has changed over time. Accordingly, it is possible to look at the empirical record for the links that connect policies to stimulate rural employment and any resulting improvement in nutritional status of the poor.[1]

---

[1] The empirical discussion of changes in rural labor productivity draws on Timmer (1989c). For a more extensive treatment of the role of food price stability in poverty alleviation, see Timmer (1989a and 1991).

Agricultural Productivity and Rural Employment

Most policy makers feel that government policy can be used to speed economic growth, improve income distribution, and provide food security to the society. In the large, heavily rural countries of Asia, all three of these goals might be served by appropriate interventions that stimulate employment growth in rural areas in such a way that real wages begin to rise. Higher real wages can be supported only by productivity growth, which contributes to greater economic output. Income distribution tends to improve when wages rise relative to profits and land rents. And food security is enhanced if, in the context of stable food prices, poor households have access to better-paying jobs.[2] It is no wonder that policy makers often focus on employment creation as the concrete task of economic policy or that they look to the agricultural sector to create most of the needed jobs. After all, agricultural workers make up more than half the labor force in every major developing country in Asia except Malaysia. Even when the nonagricultural sectors grow extremely rapidly, their base is too small to absorb all increases in the labor force.

This simple logic, and the facts that support it, lead to an equally simple question. How fast must the agricultural sector grow in order to absorb the new entrants to the labor force who cannot find jobs in other sectors? The notion that agriculture is the employer of last resort stems from conceptions of traditional agriculture rooted in dual economy models and "the moral economy of the peasant."[3] If workers remain "behind" in the agricultural sector until they are needed in the modern industrial or service sectors, the answer is fairly simple: the agricultural sector must grow enough food to provide for the rural population until workers migrate to better-paying jobs. Beyond that, planners would be concerned only with maximizing growth of industrial output.

But this simple question raises, at least implicitly, a set of deeper and more complicated issues. First, the welfare of the rural labor force also counts in a social welfare function, and this welfare extends well beyond physical quantities of food available in the countryside. One complication, then, is the desirability of raising standards of living in rural areas. Second, a subsistence standard of living for a large fraction of the population is an obvious impediment

---

[2]Higher incomes for the poor are not sufficient to improve their food security if prices of staple foods become substantially more volatile at the same time. See Timmer (1989a).

[3]The classic references are Boeke (1953) and Chayanov (1924, translated and reprinted 1966); the modern statements are in Lewis (1954), Fei and Ranis (1964), and Scott (1976).

to expanding the market for domestic manufactures. Hence a further complication is the desirability of a rapidly expanding agricultural sector to provide significant stimulus to the rest of the economy, making it easier to meet targets of industrial growth. In the face of these complexities, the question for government policy is also more complex: how fast must the agricultural sector grow to absorb the residual labor at constant or rising standards of living? Alternatively, what development strategies can raise real wages in rural areas so that the agricultural sector participates in and contributes to the overall growth process?

The answers to such questions require a broad understanding of both the supply and demand determinants of wage formation in rural labor markets. The diversity of experience in Asia is truly mind-boggling, and no single model or set of parameters can begin to capture either the static situation or dynamic behavior.[4] The purpose of this section is much more modest: to compare relative paths of labor productivity in agriculture across countries and over time and to examine alternative policy environments that might account for the striking differences observed. Increasing labor productivity is the basis for sustainable improvements in welfare; real wages cannot be increased for long without an underlying foundation of higher labor productivity. Policy makers often concentrate too much on the effect they want—higher real wages—and too little on the fundamental source of those wages—higher labor productivity. Partly because comparable data on real wages are difficult to obtain, and partly because differences in labor productivity are the important economic variables to explain, this chapter concentrates on patterns of productivity change rather than wages.[5]

Agricultural Growth and Employment in Agriculture

In most developing countries where job creation is a significant objective of the planning process, government investments and policies

[4]Binswanger and Rosenzweig (1981) summarize a large empirical and theoretical landscape that was presented at a major conference; the empirical complexities with respect to the functioning of rural labor markets and wage formation presented in the country papers at the conference clearly overwhelmed the theorists.

[5]It is recognized that raising real wages is not the same as raising labor productivity, although the two are related. Certain forms of institutional or technical change can raise average labor productivity while leaving marginal productivity unchanged or even lower. In neoclassical models of wage determination, marginal labor productivity should be equal to the wage. It is also important to stress that the wage under discussion is that prevailing in rural labor markets accessible to any individual desiring to work, not a restricted wage paid, for example, to plantation employees or workers on special government projects.

are manipulated in ways that will influence the extent of employment growth. A standard exercise is for the planning agencies in such countries to use an input-output table to calculate the number of new jobs created by the volume of investment resources available for the industrial sector. Employment in the service sector is usually assumed to be directly linked to growth in the industrial sector through some fixed multiplier. New entrants to the labor force are determined independently from population census data, and the gap between the number of new entrants available and the number of new jobs generated in the industrial and service sectors is then the "residual" number of jobs to be created by growth in the agricultural sector. The "required" rate of growth in the agricultural sector to create these jobs then depends on the elasticity of employment with respect to growth in agricultural GDP. This elasticity is a standard parameter in planning models and is usually estimated from time-series data or evidence from comparable countries.

This planning approach illustrates clearly the confusion between the role of agriculture in generating employment and the appropriate policies needed for the sector to play the assigned role. Some instrument is needed to stimulate agriculture to the necessary growth rate; the typical approach is to assume a rate of technical change that would lead to this growth rate and then to calculate the numbers of extension agents and amounts of modern inputs required to bring about the growth in output. Of course, the introduction of such technical change is likely to change the elasticity of employment from agricultural growth even if it is successfully adopted in the absence of supporting policies. But more important, the policy environment needed for the successful adoption of technical change would have significant ramifications for the rest of the economy, including the likely absorption of labor in the industrial and service sectors. The linear approach to planning the contribution of agriculture to the growth process cannot work, even when the distinction between *ex post* and *ex ante* parameters is recognized. A general-equilibrium approach that recognizes the simultaneous linkages across sectors, and the policies that influence them, is needed for simple consistency. Fortunately, the same approach also tends to highlight the type of policies that maximize the dynamic interaction between agricultural growth and growth in the rest of the economy.

The confusion between objectives in a planning model and policy instruments needed to reach the objectives becomes especially clear

Figure 5.1. Changes in the elasticity of agricultural employment from growth in agricultural GDP (E_A), from 1960–70 to 1970–80

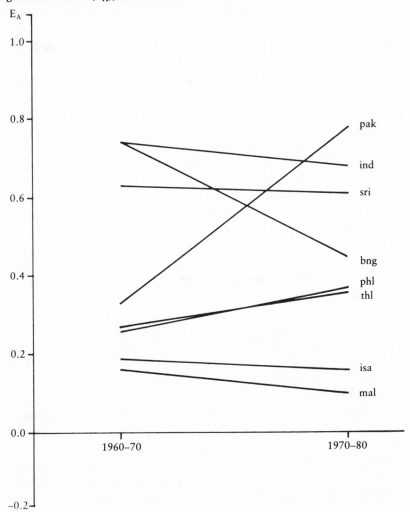

from an examination of the actual values of the employment elasticity from agricultural growth in different countries for several time periods. This empirical link between agricultural growth and employment in agriculture is simply the ratio of the two growth rates. This employment elasticity (E_A) is calculated and the results plotted in Figure 5.1 for the 1960–70 and 1970–80 periods for eight countries in South and Southeast Asia.[6] From the figure, it is immediately apparent that the elasticity of agricultural employment from growth in agricultural GDP is not a simple analytical or planning concept. In this sample, it ranges from 0.1 to 0.8. For three countries it increased between the 1960s and 1970s; for four countries it declined. One was unchanged. No obvious explanatory variables can account for the wide variation. Although the employment elasticity might be a useful summary statistic to describe growth patterns after the fact, it seems distinctly unhelpful in anticipating future growth in agricultural employment if there is little understanding of why the elasticity varies so much. Given its importance in actual planning environments, this failure of the employment elasticity to provide more reliable guidance is troubling.

A first step in gaining more insight into the relationship between the policy objective—employment growth—and the ostensible policy instrument—growth in the agricultural sector—is to plot the agricultural employment elasticity for each time period against the annual rate of growth in labor productivity for the entire work force (P) (see Figure 5.2). Labor productivity growth is defined as growth in aggregate GDP (G_Y) minus the growth in the labor force (G_L).[7] In all countries for both time periods, this measure of productivity growth is positive. The lowest level was the growth of 1.5 percent per year between 1970 and 1980 in Bangladesh; the highest was the 6.4 percent annual gain between 1960 and 1970 in Thailand. There is no clear tendency for the rate of growth in labor productivity to increase or decrease between the two periods. In Southeast Asia, it increased significantly in Malaysia, the Philippines, and Indonesia but dropped sharply in Thailand. In South Asia, it dropped sharply in Pakistan but was nearly unchanged in Sri Lanka, India, and Bangladesh.

There does seem to be a negative relationship, however, between the rate of change in aggregate labor productivity and the elasticity of employment. At one level this relationship is not surprising, because growth in labor productivity is defined to mean that less labor is needed per unit of output. The relationship shown in Figure 5.2 is more than an accounting artifact, however. The elasticity measure refers to agricultural employment and agricultural GDP, whereas the

---

[6]The country abbreviations used in Figure 5.1 and in following figures are as follows: bng = Bangladesh; ind = India; isa = Indonesia; mal = Malaysia; pak = Pakistan; phl = Philippines; sri = Sri Lanka; thl = Thailand.

[7]Thus $P = G_Y - G_L$. When additional subscripts A, I, and S are used, the relationships refer to the agricultural, industrial, and service sectors, respectively.

Figure 5.2. Elasticity of agricultural employment from growth in agricultural GDP (E_A), compared with growth in productivity of the total labor force (P), 1960–70 and 1970–80

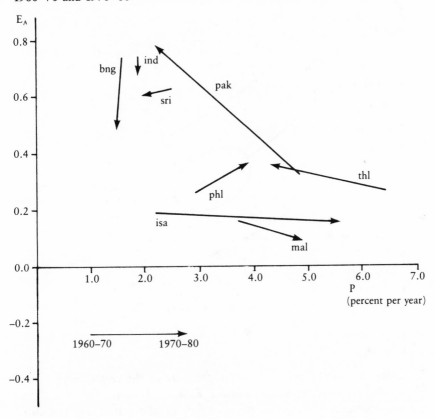

labor productivity measure is for the entire work force. From the data in Figure 5.2, two separate patterns can be identified, which are shown schematically in Figure 5.3, along with a third pattern that is demonstrated in countries with higher per capita incomes than those of these eight countries. Countries in the upper left part of the figure are primarily in South Asia and have agricultural labor forces increasing at more than half the rate of the aggregate labor force. They have relatively stagnant productivity of agricultural labor and are likely to have constant or even declining rural wages. Countries in the middle segment are mostly in Southeast Asia, have rising but significantly slower growth in the agricultural labor force relative to the total, and have rising productivity for agricultural labor. Rural wages are likely to be rising or at least about to rise if productivity gains

Figure 5.3. Stylized patterns of relationships between E_A and P

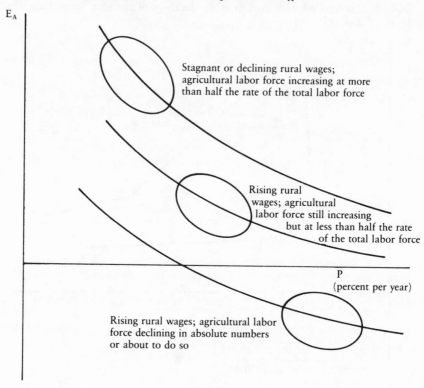

E_A

Stagnant or declining rural wages;
agricultural labor force increasing at more
than half the rate of the total labor force

Rising rural
wages; agricultural
labor force still increasing
but at less than half the rate
of the total labor force

P
(percent per year)

Rising rural wages; agricultural labor
force declining in absolute numbers
or about to do so

continue. Countries in the lower right part of the relationship, such as several in East Asia, the Middle East, and Latin America, have passed through a major turning point in the structural transformation; the absolute size of the agricultural labor force is either declining or about to decline. Significant gains in productivity of agricultural labor, when linked to rural-urban migration, imply that rural wages are rising rapidly. In this set of countries, only Malaysia is nearing this pattern.

Labor Productivity and Rural-Urban Income Distribution

From the data plotted in Figures 5.1 and 5.2, it is apparent that the elasticity of employment in the agricultural sector is an inadequate guide to prospective employment patterns. The elasticity depends fundamentally on changes in labor productivity in the rest of the economy because these changes affect rural wages, migration, and labor productivity in agriculture. These complex connections can be summarized after the fact in a single number—the elasticity of employment—but the connections themselves must be understood if a clearer picture is to emerge of the mechanisms that government policy makers can use to stimulate the role of agricultural growth in employment generation and income distribution. Various strategies can be used to increase agricultural output directly, and each has different consequences for employment. Strategic choices in the rest of the economy, especially trade orientation, also have a major impact on agriculture.

These broader concerns change the focus from trying to find a set of policies that would yield a given rate of growth in agriculture. Instead, attention is directed to understanding the relationships between trends in growth in employment and output across sectors— especially to understanding differential growth in labor productivity among the agricultural, industrial, and service sectors. An analysis of these patterns of differential growth throws considerable light on the potential for agricultural policy to play an active role in determining the extent of employment creation from increases in agricultural output. The assumption of fixed employment coefficients from growth in agriculture is not only factually wrong; a passive approach based on this notion causes substantial problems in reaching the underlying welfare objectives of the economic plan. Most important, such an approach fails to stimulate the positive linkages between the agricultural sector and the rest of the economy and thus misses the dynamic contributions from agriculture to the overall growth process.

A comparison of growth in labor productivity in the agricultural sector with growth in labor productivity for the entire economy (including agriculture) indicates whether rural-urban income distribution is likely to be improving or worsening (see Figure 5.4). Historically, the share of agriculture in GDP has always been lower than its share in the total labor force, so growth in labor productivity in the agricultural sector must be faster on average than that for the entire economy if this gross measure of rural-urban income distribution is to improve over time.[8] Such an improvement took place in the early stages of growth in most developed countries—although

---

[8]The ratio of the share of agriculture in total GDP to agriculture's share in the total labor force is only a gross measure of rural-urban income distribution because several important factors are left out of the calculations, especially remittances (which can go either direction), capital flows, the impact of investments in public goods, and rural-urban differences in amenities and cost of living. Also, as noted earlier, there is not a one-to-one correspondence between labor productivity and income in each sector even in a neoclassical world where wages equal marginal productivity. The ratio of labor productivities in the rural and urban sectors uses average values. In addition, the two-sector ratio masks potentially important distributional issues within each sector.

Figure 5.4. Relationship between growth in productivity of agricultural labor force (P_A) and growth in productivity of the total labor force (P), 1960–70 and 1970–80

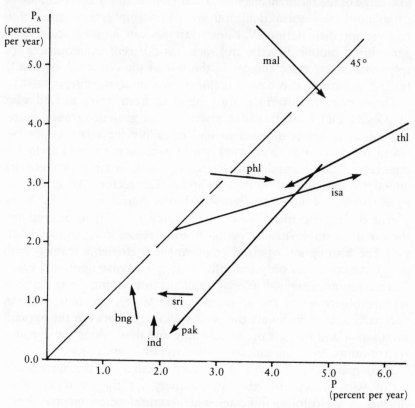

latecomers to the process, such as Italy, Japan, and the Soviet Union, showed persistent lags in labor productivity in agriculture compared with that of the industrial sector, perhaps because state investments and policy-stimulated incentives were concentrated in the industrial sector.[9] A similar lag is apparent for most of the eight countries examined here. Only Malaysia comes close to having its rates of growth in productivity for both periods above the 45° line, which shows equality in growth rates in labor productivity in agriculture and the rest of the economy.

Below the 45° line in the figure, where rural-urban income distribution continues to worsen over time, two clusters of countries are

[9]The historical patterns are described and analyzed in Kuznets (1966).

apparent. The first is the group of high-growth countries with total labor productivity growing about 4.0 to 5.5 percent per year and labor productivity in agriculture increasing at roughly half that rate. This cluster includes the Philippines, Thailand, and Indonesia for the 1970s and Pakistan in the 1960s. Thailand joined the group from "above," as its productivity performance in the 1960s established the frontier of growth for this sample. Even after a fall in rates of growth of both agricultural and overall labor productivity in the 1970s, the Thai record remained firmly in the middle of the good performers.

The record of Pakistan is startling. After a decade of impressive growth in labor productivity in both agriculture and the total economy, Pakistan rejoined the relative stagnation of South Asia in the 1970s. The sudden transition strongly suggests that something other than factor endowments and population growth rates accounts for the differences in productivity growth. The total labor force in South Asia did not grow more rapidly than in Southeast Asia in either period. Resource endowments in South Asia are not noticeably worse than those on Java (where over two-thirds of Indonesia's population lives), and yet Indonesia's productivity record fits the patterns of other fast growers rather than those of South Asia. Pakistan's good performance in the 1960s argues that policies can make a great difference to productivity growth and, by implication, to improvement in real wages and income distribution.

Growth patterns can be categorized in three ways, depending on their impact on inter- and intrasectoral income distribution with respect to agriculture. The three growth patterns are shown in Figure 5.5 as "+/+ growth," "−/+ growth," and "−/− growth." The first sign indicates impact on income distribution between the rural and urban sectors and the second sign indicates the likely impact on income distribution within the rural sector itself. The double positive label (+/+) is attached to rapid productivity growth above the 45° line. Intersectoral income distribution, at least as measured in this crude fashion, is improving simply because labor productivity is rising faster in agriculture than in the rest of the economy. The rapid growth in productivity in both parts of the economy is likely to translate into rising real wages, which should improve income distribution within the two sectors as well. In the −/+ region, by contrast, the labor force in agriculture continues to have slower growth in productivity than that in the rest of the economy, and rural-urban income distribution worsens. But agricultural and nonagricultural labor productivity is rising fast enough that real wages in both sectors

Figure 5.5. Stylized patterns of growth in labor productivity and possible paths between different types of growth

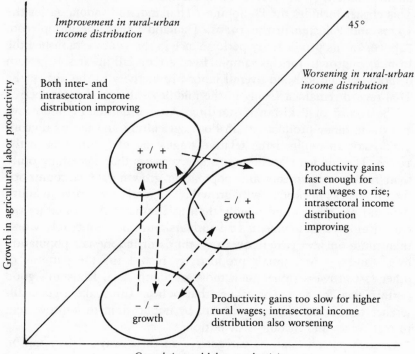

Growth in total labor productivity

are likely to be rising, thus improving income distribution within the agricultural sector. When productivity growth is very slow and is concentrated outside the agricultural sector, as in South Asia, the consequences for income distribution are likely to be doubly negative, for both inter- and intrasectoral comparisons.

Government Policy and the Path of Economic Growth

The purpose of characterizing growth in productivity by its impact on intra- and intersectoral income distribution is to highlight the consequences of different paths of economic growth and to identify any interdependencies between the pattern of growth in labor productivity and the strategy for overall growth being pursued by a country. In Figure 5.4 it is readily apparent that countries differ from one another in these growth patterns and individual countries exhibit sharply different patterns from one decade to the next. But these differences raise deeper questions. What causes a country to be in one region rather than another, and what policies lead to shifts in regime? Can government policies alter the speed by which countries traverse the transition paths from one region to another? All countries would like to have more rapid growth in labor productivity in the agricultural than in the nonagricultural sectors in order to improve rural-urban income distribution, with labor productivity in both sectors rising fast enough to cause wages to rise. That is, all countries would like to have +/+ growth and to formulate a policy to reach it.

Figure 5.5 plots the six possible transition paths from one region to another, but Figure 5.4 reveals that not all possible growth paths have historical precedents in this sample. The road most traveled runs in the wrong direction, from +/+ to −/+ growth. Malaysia, the Philippines, and Indonesia took this route from the 1960s to the 1970s. Thailand seemed headed the same way. Pakistan, as already noted, fell from −/+ to −/− growth. Once in the −/− growth pattern, no country escaped in the 1970s, which suggests that either political or economic hysteresis is an important factor in explaining the poor performance of South Asia. Important changes in policy made in the late 1970s and the 1980s, however, do not show up in the empirical record depicted in Figure 5.4. Bangladesh and Sri Lanka are pointed in directions that suggest they might join the +/+ cluster. Agricultural GDP in both countries grew substantially more rapidly between 1980 and 1986 than in the prior two decades; there is the possibility of a steep rise in labor productivity if the agricultural labor force grows less rapidly than the total labor force. Impressive rises in industrial and services GDP in both countries make this quite likely.

Policies to Maintain Growth in Agricultural Productivity

The movement from +/+ growth to −/+ growth, where growth in productivity for agricultural labor lags behind that for nonagricultural labor but both rates remain relatively high (above 3.0 percent per year), is not difficult to explain for those Asian countries in which it occurred. Malaysia and Indonesia suffered to some degree from "Dutch Disease," the relative loss in competitiveness for the tradable-goods sector during an export boom in natural resources, in this case petroleum, affecting the agricultural sector in the 1970s. The significant feature in both countries was the maintenance or improvement in rates of growth in productivity of agricultural labor, while the rest of the economy was stimulated to even greater gains by oil dollars. Part of the reason was the oil stimulus itself and the impact of a booming construction sector on rural wages. But for Malaysia and Indonesia, the government's continued concern for welfare in the rural sector was part of the story. Both governments maintained the competitiveness of traditional agricultural exports through careful macroeconomic management, either through control over inflation (Malaysia) or competitiveness of the real exchange rate (Indonesia). The important role of macroeconomic policy as a key ingredient in the good performance of the agricultural sectors in both countries is now recognized as a textbook case of the interdependence of the two areas of the economy.[10]

Because so many countries are concentrated in the $-/+$ region, fairly basic economic forces must be at work to generate this pattern of growth. Consequently, it is important to know how stable the growth pattern is and where an economy ends up after growing in this fashion for some time. The major economic success stories of East Asia (Japan, Taiwan, and Korea) followed the $-/+$ growth pattern for several decades. All ended with severe cases of structural lag. Too many resources were left in agriculture as the industrial economy spurted ahead, and these societies encountered major domestic and international political problems because of the price and trade policies used to raise farmers' incomes and protect the agricultural sector from foreign competition. Of the countries beginning in the 1980s in the $-/+$ region, only Thailand did not use similar trade and pricing policies for key commodities in an effort to protect domestic farmers from the very low prices that occur from time to time in world markets. Although the strong performance of Thailand in terms of rising labor productivity argues that such free-trade policies promote growth, Thailand paid a price in terms of rural poverty. Other countries in the region have had excellent growth records but pursued different policies.[11]

Sectoral Patterns of Productivity Growth

A successful $-/+$ growth path is one that maintains significant growth in productivity of agricultural labor, even though start-up in-

---

[10]See the discussion in Gillis et al. (1983) of Dutch Disease in Indonesia as an illustration.

[11]See Timmer (1989b) for a discussion of comparative rural poverty. See also the summary of the country papers in Sicular (1989), pp. 289–95, for a judgment on the role of pricing policies in agricultural and overall economic growth.

dustrialization through import substitution introduces powerful biases against agriculture. Consequently, for a country to stay on the −/+ path requires compensatory investment and pricing policies to maintain acceptable rates of productivity growth in agriculture. A policy of import substitution is normally implemented through substantial, even prohibitive, trade barriers for competitive products, and these barriers have been attacked by economists and donor agencies as allocatively inefficient and prone to capture by vested interests. Such barriers may be inefficient only in a short-run, static sense, however. The important issue, at least from an East Asian perspective, is whether subsequent policies of trade liberalization and other efficiency-enhancing measures at the macroeconomic level can convert the industrial base and "learning by doing" created by import substitution into export competitiveness. Such policy measures tend to promote labor-absorbing investments, thus speeding the economy toward the turning point at which real wages start to rise. With policies that favor integration of urban and rural labor markets and widespread rural industrialization, trade liberalization and stable macroeconomic policies propel the economy toward a pattern of +/+ growth.

If growth of the economy is as rapid as that of East Asia since the 1960s, a serious problem of structural imbalance is likely to result. Even with policies that favor rural industrialization and easy migration to urban jobs, the sizable investment in fixed agricultural capital is likely to keep more labor in the agricultural sector than is commensurate with labor productivity rising more rapidly in the agricultural sector than in the rest of the economy. If the agricultural sector becomes heavily protected at this stage, as it did in East Asia, the transition will be slowed and a pattern of −/+ growth will emerge. It is important to recognize, however, that such protection may be essential for the domestic political stability that provides confidence to investors to take the long view with respect to risky investments.[12]

The Fall from −/+ to −/− Growth. A fall from fairly rapid −/+ growth to much slower −/− growth is never an objective of government policy makers, but its occurrence nonetheless is explained by the complex and poorly understood links between industrial and agricultural policies. A capital-intensive spurt of import-substituting

---

[12]See Anderson and Hayami (1986) and Reich, Endo, and Timmer (1986) for further discussion of the political economy of agricultural pricing in East Asia, its relevance to the rapidly growing economies of Southeast Asia, and the role of agricultural trading partners, especially the United States, in pressing for changes in domestic agricultural policies.

industrialization behind high trade barriers raises labor productivity in the industrial sector (although total factor productivity is usually low). Labor productivity in the service sector might also increase if government expands rapidly to manage and implement the strategy and pays high wages to its new employees performing these tasks.[13] With labor productivity rising rapidly in the nonagricultural sector, the country experiences −/+ growth if labor productivity in the agricultural sector does not lag too badly. In the early stages of an import-substitution strategy, momentum from previous investments in the agricultural sector and outward-oriented trade policies continues to raise agricultural labor productivity. But the strategy is eventually self-defeating if there is no liberalization of industrial policy and a conversion of the economy to export competitiveness. A failure in this regard is likely to send even large economies into a pattern of −/− growth. This pattern of tightened inward orientation instead of conversion to export competitiveness seems to have been the fate of Pakistan between the 1960s and the 1970s, and Thailand may have followed this path in the absence of reforms in industrial policy that stimulated its growth in nonagricultural exports in the 1980s. The near collapse of the Philippine economy after impressive growth in productivity in the 1960s and 1970s can also be partially attributed to a failure to reform industrial policy, although the highly extractive nature of both industrial and agricultural policy that was visible by the early 1980s may be adequate explanation by itself.

This is a bleak judgment. Not only does an inward-oriented industrial (and service sector) policy lead to slow growth in productivity in the industrial sector after the initial spurt, it eventually spills over to the agricultural sector as well and causes slow growth in productivity there. In reverse, however, the linkages can be positive. When policies consciously seek to connect rural and labor markets through better communications and transportation systems and stimulate labor absorption in the industrial sector through appropriate macro prices, labor productivity in agriculture also rises. Equally important, the dynamics transmitted to the agricultural sector reinforce those in the industrial and service sectors through pressures on real wages to rise.

Stimulating Intersectoral Linkages. The potential to use policies for the industrial and service sectors to stimulate productivity growth

---

[13]Labor productivity in the government sector is *measured* by the wage bill, so the increase is, to a large extent, by definition.

in agriculture can be examined by plotting the growth rates in labor productivity for the three sectors on a back-to back diagram. Patterns for Southeast Asia and South Asia are shown in Figures 5.6 and 5.7, respectively. Growth in agricultural productivity, shown on the vertical axis, increased in Indonesia and decreased in Malaysia, Thailand and, marginally, in the Philippines. Each country in Southeast Asia shows a striking symmetry of rates of change in productivity between the industrial and service sectors (see Figure 5.6). For both increasing and decreasing rates of productivity growth in agriculture, the figure shows that whenever the rate increased in the industrial sector between the 1960s and the 1970s, it also increased in the service sector, usually by similar magnitude. Identical symmetry also holds between the service and industrial sectors in South Asia (Figure 5.7), but the entire distribution of changes is shifted downward, compared with that of Southeast Asia, because of lower rates of growth in productivity for agricultural labor in South Asia. The regional clusters relative to the 45° line of equal productivity growth in the respective sectors show this shift quite dramatically.

In Southeast Asia, patterns of productivity growth in both the service and industrial sectors cluster around the 45° line; industrial productivity tends to grow slightly faster than agricultural productivity and service productivity slightly slower. But the balance among the three sectors is striking, which suggests that integration of labor markets across the three employment fields is reasonably good.

South Asia, by contrast, shows a uniform pattern of higher productivity growth for labor employed in the service and industrial sectors than that for agricultural workers. The exception is the steep drop in labor productivity in the service sector in Bangladesh for the 1970s.[14] The South Asian patterns suggest that a significant dualism in labor markets still exists, and agriculture remains the employer of last resort. Again, Bangladesh seems to be an exception, but not in a positive way. The pattern of low or negative productivity growth in the service sector suggests that it has become the residual employer; agriculture is no longer able to accept more workers, even at extremely

[14]The Bangladesh data are especially questionable. Figure 5.7 shows an alternative calculation for Bangladesh in dashed lines. This alternative is based on a lower share of the labor force in agriculture in 1970 than is reported in the data source used for all other data on labor force shares for 1960 and 1970, i.e., *World Tables, 1980*, published by the World Bank. The alternative data for Bangladesh for 1979 come from the 1987 edition of the same publication, and it is the only country in this sample with such major differences. Unfortunately, the 1987 edition does not report labor force shares for industry and services.

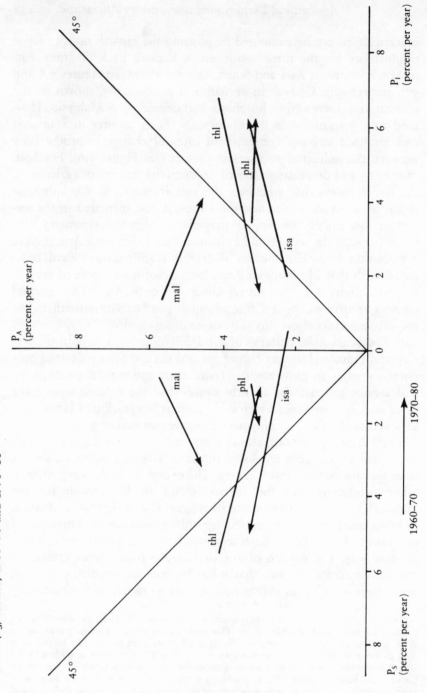

Figure 5.6. Relationships in Southeast Asia among growth in labor productivity in the agricultural (P_A), industrial (P_I), and service (P_S) sectors, 1960–70 and 1970–80

Figure 5.7. Relationships in South Asia among growth in labor productivity in the agricultural (P_A), industrial (P_I), and service (P_S) sectors, 1960–70 and 1970–80

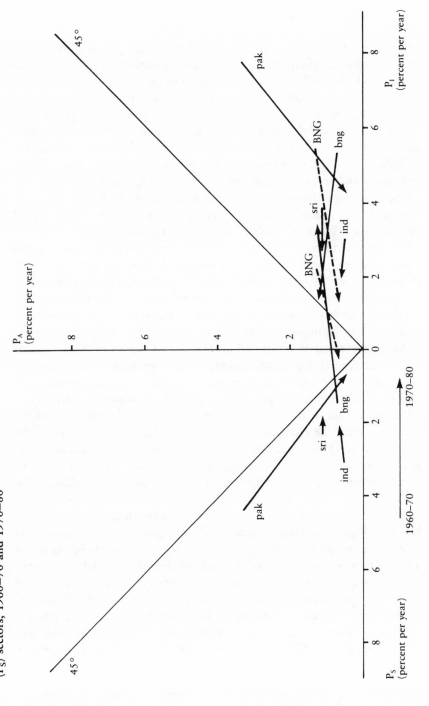

low wages. Changes in caloric availability during this period, discussed below, confirm this negative interpretation of the changes in labor productivity in agriculture and services.

Despite the contrasts in levels of productivity growth in agriculture between Southeast and South Asia, the symmetry of responses between the industrial and service sectors argues that some common mechanisms are at work in the labor market. This strong symmetry is difficult to explain, however. Changes in labor productivity in the industrial and service sectors were similar in sign and magnitude, no matter what was happening to labor productivity in agriculture.[15] In Southeast Asia, rates of growth in labor productivity in the industrial and service sectors fell only in Thailand, where the growth rate in labor productivity in agriculture was also dropping between the 1960s and the 1970s. In the other three countries, the rate of growth in productivity in both the industrial and service sectors increased between the two periods. In Indonesia the rate of growth in agricultural productivity was also rising, but in Malaysia it fell significantly and in the Philippines slightly.

No explanation of integrated labor markets works by itself, but for Southeast Asia this general pattern, in combination with unique aspects of each country's experience, probably provides an adequate explanation. Indonesia's recovery in growth of labor productivity in all three sectors in the 1970s is thus due to the rehabilitation of the economy after the chaos of the 1960s, and the relatively modest improvement in labor productivity in agriculture must be attributed to Dutch Disease. Similar macroeconomic problems explain the decrease in rate of productivity growth in agriculture in Malaysia, but policies that favored rural areas, an export orientation for agricultural products, and continued agricultural investments kept the absolute rate high. Thailand suffered a sharp drop in labor productivity growth across all three sectors in the 1970s because of the oil shock and the failure of the industrial sector to absorb much labor. Only in the 1980s did the restructuring of the Thai economy seem to offer much stimulus to industrial labor absorption. By the mid-1980s commodity prices in the world market were depressed, however, and the substantial numbers of workers remaining in agriculture had very low incomes. The Philippines maintained a reasonable rate of growth

[15]The alternative data for Bangladesh do not show the same symmetrical pattern.

in labor productivity in agriculture in both periods.[16] But the acceleration in industrial productivity came primarily through inefficient import substitution, and growth in labor productivity in the service sector remained quite low, which suggests that it was the service sector rather than agriculture that was beginning to serve as the employer of last resort.

Compared with Southeast Asia, the South Asian patterns would look very similar if labor productivity in agriculture were rising at roughly 3 percent per year instead of the observed level of less than 1 percent. But even if agriculture could be stimulated independently of the other two sectors, the arrows point the wrong way for changes in rates of growth in labor productivity in industry and services between the 1960s and the 1970s. The rates fell for all four countries in both sectors. Furthermore, the rates of productivity growth in the service sector are very low, even negative, which suggests a substantial push of laborers out of agriculture into the informal service economy. In the 1970s, the industrial sector seems not to have been the direct cause of this push, because growth in labor productivity in the industrial work force was also quite low, which suggests that a considerable number of laborers was being absorbed relative to output growth. All of the economies of South Asia apparently reached a low-level equilibrium in the 1970s, possibly because of high oil prices and imports of expensive grain in combination with inefficient, protected industrial sectors (but not ones that were excessively capital-intensive). Restructuring economies with such deep problems in all three sectors involves massive changes in policy and considerable disruption to "business as usual" in both government and business communities. Even by the late 1980s, it was not clear that South Asia had found the right approaches and combinations of policies to accomplish this restructuring successfully.

Part of the difference in patterns of growth in labor productivity between South and Southeast Asia may be accounted for by a different mix of public and private investment in rural areas. Southeast Asia has uniformly relied more heavily than South Asia on the private

---

[16]Micro data collected by the International Rice Research Institute (IRRI) on real wages by specific agricultural task show a significant *decline* during this period, despite the increase in labor productivity shown by the aggregate data. A logically consistent reconciliation of the two trends requires that owners of land and rural capital captured the incomes from the higher labor productivity, but no specific evidence is available to test the hypothesis. The data are shown in Ahmed (1988).

sector to market basic food grains, build rural infrastructure, and invest in and manage delivery systems for productivity-enhancing inputs. The results of these private sector activities are fully reflected in their contribution to agricultural GDP. A partially compensating factor in South Asia is the greater reliance on the public sector for these and other activities; a higher share of real income in rural areas is thus reflected as consumption of public goods, which does not show up in the national income accounts. The compensation is only partial, however, because eventually labor productivity must rise in all sectors to support a rising stream of investment, be it public or private.

Agricultural Policy and Welfare of the Poor

The main cause for concern over lagging labor productivity in any sector of the economy, but especially in agriculture, is its implication for the standard of living of poor people. Comparative data on standards of living are notoriously difficult to collect and evaluate, but trends in labor productivity by sector reflect the underlying economic environment in which these people live and work. The link between labor productivity and standards of living is not rigid, of course, because wages and prices are intervening variables. But living standards of the poor should rise if food prices are stable and if increased opportunities for employment cause wages to rise. A measurable proxy for improvement in income distribution and welfare of the poor is improvement in their caloric intake, especially if the starting point was well below recommended levels. In this context, labor productivity is an essential support for sustained access to goods and services that come from the market. For the rural poor in all the countries in this sample, most food supplies come from the market or their own farms despite the prevalence of food rationing programs and fair price shops in South Asia. Consequently, labor productivity and the incomes supported by it, along with the prices of food in the markets, are the major determinants of food intake and the welfare of the rural population. By looking at changes in food consumption over time, it is possible to judge, however roughly, the impact of government policies to raise labor productivity and to stabilize food prices.

Indicators of Improving Welfare

The distribution of food consumption, especially caloric intake, has often served as a proxy for the broader measure of distribution of income, since income distribution is hard to measure even at a single point in time and is doubly hard to track over time. Household food consumption surveys are frequently repeated at five- or ten-year intervals using similar protocols and sample frames, so reasonable inferences can be made about distributional changes over time. On a more immediate basis, changes in average caloric intake in a country offer substantial insight into changes in income distribution over time, and such data are available on an annual basis for most countries. Comparisons across countries and over time offer a relatively quick and easy approach to the analysis of comparative patterns of income distribution, or at least one important component of it for which policy makers express concern.

Average caloric intake level for a particular year and country is correctly criticized as a welfare indicator because the distribution of levels around the average is not discernible from the average itself. But when the average for the entire society changes significantly over time, substantial implications for welfare change are implied. Middle- and upper income households have very low income elasticities of demand for calories. If average caloric intake increases or decreases from year to year, most of the changes are due to altered caloric intake in poorer households.[17] When a country increases its average daily per capita intake of calories from well below the recommended average to well above it, the only explanation is that low-income households are better fed. Stagnation or deterioration in this measure means a lower standard of living for the poor.

Table 5.1 presents the basic data to examine trends in caloric intake for the eight Asian countries in the sample for the period from 1965 to 1985. The diversity is substantial. Daily calorie supplies available, the nearest available proxy for intake, ranged from a low of 1,747 kilocalories (kcal) in Pakistan in 1965 to 2,684 kcal in Malaysia in 1985. Relative to recommended levels of intake, based on age structure, activity levels, and climate, Pakistan's intake in 1965 was nearly 25 percent too low, whereas Malaysia's 1985 intake was 20 percent above average recommended levels. Despite substantial disagreement over the true welfare significance and validity of recommended nutritional levels on average, they do provide a useful benchmark that is corrected for the most important differences in population structures and nutritional needs. Any country with average

---

[17]The decline in income elasticity of demand for calories as incomes increase reflects the interaction of Engel's, Bennett's, and Houthakker's laws. See chapter 2 of Timmer, Falcon, and Pearson (1983) for a fuller discussion and references to the empirical literature.

Table 5.1. Changes in caloric availability in representative countries in Asia, 1965–85

Region and country	Daily calorie supply			Supply as percent above or below average calorie requirements		
	1965	1985	Annual % change	Level[a]	1965	1985
Southeast Asia						
Malaysia	2,249	2,684	0.9	2,232	0.8	20.2
Thailand	2,200	2,462	0.6	2,219	−0.9	11.0
Philippines	1,936	2,341	1.0	2,266	−14.6	3.3
Indonesia	1,742	2,533	1.7	2,164	−17.2	17.1
South Asia						
Pakistan	1,747	2,159	1.1	2,320	−24.7	−7.0
Sri Lanka	2,155	2,385	0.5	2,215	−2.7	7.7
India	2,100	2,189	0.2	2,200	−4.7	−0.6
Bangladesh	1,964	1,899	−0.2	2,300	−14.6	−17.4

Source: Data from World Bank, *World Development Report, 1987* (New York: Oxford University Press for the World Bank, 1987).
[a]Based on 1983 population structure.

caloric intake significantly below the recommended level almost inevitably has a sizable proportion of the population, usually in rural areas, that would like to consume greater quantities of food if their income levels permitted. This connection to incomes of the poor allows changes in caloric intake over time to be used as a rough proxy for changes in welfare levels of the poor even in the absence of direct statistics on income distribution.

In 1965, only Malaysia had levels of average caloric intake at or above such recommended levels; the unweighted average deficit was 9.8 percent. By 1985, only Pakistan, India, and Bangladesh remained below recommended levels, and the unweighted average surplus was 4.3 percent. On average, the Asian region improved its per capita caloric intake, relative to needs, by 14.1 percent, from well below to somewhat above recommended levels—all in two decades. The improvement is most dramatic in Southeast Asia. South Asia's gains were much more modest, and Bangladesh in fact slipped backward.

Changing Patterns of Food Consumption

Explanations for the changes in caloric intake across the eight countries are more complicated than might be expected. Figure 5.8 plots the average annual percentage change in per capita caloric intake (CGAIN) against growth in average per capita incomes (YAVG). A

Figure 5.8. Relationship between the increase in per capita calorie intake (CGAIN) and average per capita income (YAVG), 1965–85

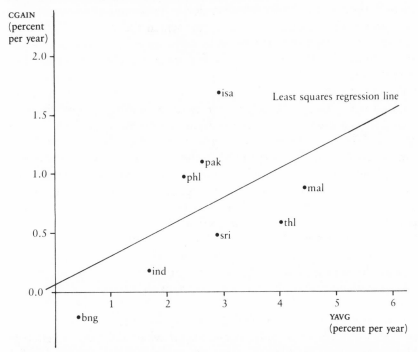

rough positive relationship is apparent, but the income variable leaves substantial variance unexplained in a simple regression. A more satisfactory model includes the size of the initial gap between recommended and actual intake levels (GAP) as a second explanatory variable along with the gain in average per capita income for the entire population. But this regression, after correcting for degrees of freedom used in the regression, explains only two-thirds of the variance in the growth of per capita caloric intake for the eight countries between 1965 and 1985.

Factors other than the size of the initial caloric deficit and growth in total incomes are important for explaining the changes in average caloric intake. Changes in agricultural incomes, income distribution, and food prices are likely to be the key omitted variables. But that is precisely the point. As Figure 5.8 shows, the main outliers in the regression analysis are Indonesia on the positive side and Thailand on the negative. India, Bangladesh, and Sri Lanka are also uniformly below the regression line, but not by a great amount. The rapid growth in caloric intake in Indonesia is partly accounted for by the recovery in the economy after 1965, not all of which is captured in per capita income figures. But Indonesia also devoted substantial resources to a successful price-stabilization program (which did not subsidize consumers on average over the period), and this effort, plus rapidly rising production of rice, accounts for much of the nutritional improvement.

Thailand's slow gain in caloric intake relative to its growth in per capita income can be accounted for by deteriorating income distribution between the rural and urban areas during the second half of the period. World prices for most agricultural commodities that Thailand exports were very depressed in the mid-1980s. These low prices caused agricultural incomes to grow much less rapidly than the growth in labor productivity in the agricultural sector. In equations where growth in agricultural income enters the regression independently, Thailand's low growth in caloric intake is no longer an outlier.

The three negative deviations in South Asia, although not substantial, are important because of the regional pattern. South Asia has already been singled out as having low growth in labor productivity, low growth in per capita income, and a likely deterioration in rural wages. The data for caloric intake support this characterization. Sri Lanka grew fairly rapidly during the second part of the 1965–85 period, but with noticeable worsening of what had been a remarkably even income distribution. Average caloric intake increased in Sri Lanka, but not as much as it would have if the previous distribution of income had been maintained. More troubling perhaps is evidence of a deterioration in the food intake of the bottom income decile during the period of most rapid growth.[18]

India and Bangladesh had very little growth in income or productivity, and average caloric intake was virtually stagnant. On the basis of the parameters for the rest of the sample, however, this limited growth in caloric intake was less than would be expected. The rural income distribution deteriorated as real wages fell. Higher food prices in India succeeded in stimulating greater production, but they did little to improve the food intake of the bottom 40 percent of the population thought to suffer caloric deficits (Reutlinger and van Holst Pellekaan). The worsening distribution of land in Bangladesh, a result of population pressure and frequent crop failures that forced small landholders to sell, in combination with only limited increases

[18]See Sahn (1988) for further discussion of the new growth strategy in Sri Lanka after 1978 and its impact on income distribution and caloric intake by income class.

in demand for landless laborers, has exacerbated the situation of the poor in that country. The 1974–75 famine also seems to have permanently reduced the demand for agricultural labor after the massive migrations in search of food and jobs.[19]

The results from analyzing changing patterns of caloric intake support the emphasis in this chapter on understanding the appropriate role of government interventions in policies for the agricultural sector that are designed to stimulate higher labor productivity and increases in real wages.[20] The countries in this sample that have been successful—Malaysia and to some extent Indonesia—have seen rapid improvements in the distribution of food consumption. Countries that lagged in productivity growth, especially those in South Asia, failed to improve food intake significantly, despite efforts to stabilize food prices and make food more accessible to the poor in urban ration shops.

Strategies and Prospects

The most powerful lessons on the relationship between agricultural change and income distribution are simply replays of the dominant themes of this chapter: the need for a development policy to stimulate agricultural productivity and to foster the intersectoral links that contribute directly to agricultural development, employment, and rising real wages. When the industrial and service sectors are growing efficiently and have strong market linkages to the rural economy, an agricultural sector that grows fast enough to raise labor productivity (combined with a price-stabilization policy that assures income gains to farmers and access to food for low-income consumers) can raise rural wages and improve income distribution. There are no tricks here; only a coherent food and agricultural policy maintained for several decades can make a sustainable difference to the poor.

The diversity of experience within Asia in the past several decades offers important opportunities for comparative policy analysis. Several countries seem poised for rapid structural change and improvements in employment and income distribution. Several seem bogged down in low-growth patterns that offer little hope of significantly higher incomes. Several may well be headed backward; a deterioration

---

[19]See Ravallion (1987) and the discussion of Bangladesh in Ahmed (1988).
[20]The importance of stabilizing food prices is stressed in Timmer (1989a and 1991).

of living standards is likely unless major changes in policy lead to economic restructuring.

In the specific context of each country, two basic forces are at work to produce these results. Rising labor productivity in agriculture is a necessary but not sufficient element for rural wages to increase over time. In a related fashion, higher incomes are the basic guarantee of sustained improvements in the diets of the poor, despite the increases in per capita food consumption that can be stimulated temporarily by short-run measures.[21] The chain of causation is labor productivity to wages to income distribution. Policies can reinforce each link in the chain in such a way that improvements in income distribution are amplified or dampened. Trickle-down growth benefits the poor only when the links are amplified by specific governmental design; in market-oriented economies, without effective government intervention, such benefits can be negligible or even negative.

What does the experience of eight Asian countries since the 1960s tell us about effective government interventions to stimulate growth in employment, improvements in income distribution, and the role of agriculture in the process? If raising labor productivity in agriculture is taken as the essential starting point, two complementary paths are open. The first approach concentrates on raising agricultural output through the package of investments, new technology, and incentives, an approach that is well understood, at least in principle, throughout Asia (Hayami and Ruttan, 1985; Timmer, 1988; Ahmad, Falcon, and Timmer, 1989). The second approach concentrates on raising the labor intensity of the modern industrial and service sectors through more appropriate choices of techniques and products. Evidence that has accumulated since the late 1960s demonstrates that import substitution behind highly protective trade barriers creates perverse incentives for investment with respect to both choices (Pack, 1988). Low labor absorption and inefficient use of capital in the industrial sector prolong the dualistic nature of the labor market. Many workers with extremely low productivity are left behind in agriculture or, increasingly, in an informal service sector in which real wages are extremely low and uncertain.

Such dualism significantly exacerbates the task of integrating rural and urban labor markets in a manner that draws labor out of agriculture to more productive jobs in the nonagricultural economy. An

[21]See Lal (1985), especially the discussion of basic needs on pages 100–102, for a forceful exposition of the role of productive employment in guaranteeing the sustainability of consumption gains.

alternative path to higher labor productivity in agriculture is more rapid emigration from the sector. In this sample, net migration varied from nil in Bangladesh and India, where growth in agricultural labor force was nearly as large as for the entire labor force, to a rate of out-migration that left the agricultural labor force growing very slowly, as in Malaysia and Indonesia. No country has been able to move labor out of agriculture fast enough to prevent lower absolute productivity of labor in agriculture than that in the rest of the economy.[22] When the process is faster rather than slower, however, it contributes to higher labor productivity in agriculture and, via the supply side in rural labor markets, to higher real wages.

Most government policies that influence rural wage formation through an impact on agriculture do so via the demand side of the equation. Governments have several main instruments available to them. They can invest in rural infrastructure, including irrigation with its second-round effect on multiple cropping. They can promote new technologies that raise yields, increase labor requirements, shorten the growing season, and permit a second or third crop. They can provide adequate price incentives to stimulate on-farm savings and investments and roundabout expenditure multipliers. They can create a favorable environment for vertical diversification, which steadily transfers workers from agriculture to industry and the service sector, even if it leaves them in rural areas and living on the farm. These are the ingredients of agricultural development and structural change. Policies to reduce poverty in agriculture are largely synonymous with policies that stimulate employment and rising wages in other sectors of the economy. Their successful implementation depends on a healthy relationship, in terms of both market linkages and policy balance, between the agricultural sector and the rest of the economy.

References

Ahmad, Junaid Kamal, Walter P. Falcon, and C. Peter Timmer. 1989. "Fertilizer Policy for the 1990s." HIID Development Discussion Paper 293. Cambridge: Harvard Institute for International Development.

Ahmed, Rais uddin. 1988. "Employment and Income Growth in Asia: Some Strategic Issues." Prepared for the USAID/ANE conference Agriculture in the

[22]See Bellerby (1956) for an analysis of the chronic problems of low incomes in agriculture faced by a number of Western countries during their development process.

1990s, Washington, D.C., September 1988. Washington, D.C.: International Food Policy Research Institute. Typescript.

Anderson, Kym, and Yujiro Hayami, with associates. 1986. *The Political Economy of Agricultural Protection: East Asia in International Perspective*. London: Allen and Unwin.

Bellerby, J. R., with others. 1956. *Agriculture and Industry Relative Income*. London: Macmillan.

Binswanger, Hans P., and Mark R. Rosenzweig. 1981. *Contractual Arrangements, Employment and Wages in Rural Labor Markets: A Critical Review*. New York: Agricultural Development Council and International Crops Research Institute for the Semi-Arid Tropics, India.

Boeke, J. H. 1953. *Economics and Economic Policy of Dual Societies as Exemplified by Indonesia*. New York: Institute of Pacific Relations.

Chayanov, A. V. 1924. *On the Theory of the Peasant Economy*. Reprinted 1966 and edited by D. Thorner, B. Kerblay, and R. E. F. Smith. Homewood, Ill.: Richard D. Irwin.

Fei, John C. H., and Gustav Ranis. 1964. *Development of the Labor Surplus Economy: Theory and Policy*. Homewood, Ill. Richard D. Irwin.

Gillis, Malcolm, Dwight H. Perkins, Michael Roemer, and Donald R. Snodgrass. 1983. *Economics of Development*. New York: Norton.

Hayami, Yujiro, and Vernon W. Ruttan. 1985. *Agricultural Development: An International Perspective*. 2d ed. Baltimore: Johns Hopkins University Press.

Kuznets, Simon. 1966. *Modern Economic Growth*. New Haven: Yale University Press.

Lal, Deepak. 1985. *The Poverty of Development Economics*. Cambridge: Harvard University Press.

Lewis, W. Arthur. 1954. "Economic Development with Unlimited Supplies of Labor." *Manchester School* 22 (May): 139–91.

Pack, Howard. 1988 "Industrialization and Trade." In Hollis B. Chenery and T. N. Srinivasan, eds., *Handbook of Development Economics*, vol. 1, pp. 333–80. Amsterdam: North-Holland.

Ravallion, Martin. 1987. *Markets and Famines*. New York: Clarendon Press for Oxford University Press.

Reich, Michael R., Yasuo Endo, and C. Peter Timmer. 1986. "Agriculture: The Political Economy of Structural Change." In Thomas K. McCraw, ed., *America versus Japan*, pp. 151–92. Boston: Harvard Business School Press.

Reutlinger, Shlomo, and Jack van Holst Pellekaan. 1986. *Poverty and Hunger: A World Bank Policy Study*. Washington, D.C.: World Bank.

Sahn, David E. 1988. "The Effect of Price and Income Changes on Food-Energy Intake in Sri Lanka." *Economic Development and Cultural Change* 36, no. 2: 315–40.

Scott, James C. 1976. *The Moral Economy of the Peasant.* New Haven: Yale University Press.

Sicular, Terry, ed. 1989. *Food Price Policy in Asia: A Comparative Study.* Ithaca: Cornell University Press.

Timmer, C. Peter. 1988. "The Agricultural Transformation." In Hollis Chenery and T. N. Srinivasan, eds. *Handbook of Development Economics,* vol. 1, pp. 275–333. Amsterdam: North-Holland.

———. 1989a. "Food Price Policy: The Rationale for Government Intervention." *Food Policy* 14 (February): 17–27.

———. 1989b. "Agriculture and Structural Change: Policy Implications of Diversification in Asia and the Near East." HIID Development Discussion Paper 291. Cambridge: Harvard Institute for International Development, June.

———. 1989c. "The Role of Agriculture in Employment Generation and Income Distribution in Asia and the Near East." HIID Development Discussion Paper 292. Cambridge: Harvard Institute for International Development, June.

———. 1991. "Food Price Stabilization: Rationale, Design, and Implementation." In Dwight H. Perkins and Michael Roemer, eds., *Reforming Economic Systems.* Cambridge: Harvard University Press.

Timmer, C. Peter, Walter P. Falcon, and Scott R. Pearson. 1983. *Food Policy Analysis.* Baltimore: Johns Hopkins University Press for the World Bank.

World Bank. 1980. *World Tables, 1980.* Baltimore: Johns Hopkins University Press for the World Bank.

6

Government-Sponsored
Rural Development:
Experience of the World Bank

Graham Donaldson

Rural development has been one of the main intervention strategies used since the mid-1960s by governments of developing countries and international development agencies to raise agricultural
productivity and improve the quality of rural life. It was the single
most important element of the development strategy adopted by the
World Bank in the early 1970s, and it remains the central focus of
the Bank's current agricultural development program.

As the term implies, rural development is concerned with promoting change toward higher levels of productivity, consumption, welfare, and social organization for those who find their livelihood in
rural areas. Since most people in rural areas of developing countries
are poor, rural development is concerned with improving the productivity of the rural poor. This approach at once distinguishes it from
"basic needs" approaches which focus on improving the welfare of
the poor through improved social services, without the essential
change in productivity. Although growth in productivity may follow
from a basic needs program, it is central to rural development. Nevertheless, rural development has a social dimension because of its focus on alleviation of poverty. Since the majority of families in rural
areas are subsistence-oriented farmers, or smallholders, the focus is
largely on smallholder development.

Rural development thus constitutes a particular approach to agricultural development, one which, unlike typical commodity-oriented

programs, pays more attention to organization, institutional development, and social factors. It provides an alternative to the technology-led, large-scale agricultural development experience of Western countries in the twentieth century.

Increasing the productivity of smallholders necessitates augmenting their land by the use of purchased inputs, such as fertilizer and improved seed, or their labor by the use of better equipment or additional services. As production expands, individual producers often rapidly increase their salable surplus of farm produce. The acquisition of inputs or the sale of output both require commercial exchanges. Rural development is thus inevitably concerned with the commercial development of agriculture and with the linkages and institutional arrangements that are essential to it.

Further, since the individual transactions are typically small, arrangements are needed to aggregate salable surpluses or break down and distribute purchased supplies. This process calls for the growth of some internal specialization within villages or the introduction of external agents who will provide that service. Either way, it introduces more dealings with the broader economy, including merchants, middlemen and traders, banks and government regulators, and purveyors of services. Dealing with these various individuals or entities also requires internal organization within villages or groups. Hence rural development is concerned with the formation and evolution of local-level organizations and institutions.

The elements of rural development as outlined above are hardly unique or original. They describe essentially the same processes observable in history. For instance, these dynamics were evident as feudal Europe moved out of the Dark Ages. In the eleventh century this transition was described in relation to the emergence of market towns such as Frankfurt, Augsburg, and Champagne, which provided new market opportunities beyond the villages and marked the beginning of commercial development in Europe. But the process was slow, without the support of benevolent government services. More recently, a similar path has been followed in other countries, including Japan, Taiwan, Indian Punjab, Korea, and peninsular Malaysia. These more recent examples, however, have been government-supported endeavors.

Government-supported rural development is a phenomenon of the twentieth century whereby governments have sought, through various interventions, to speed up the processes of commercial and social development overall or in certain less prosperous areas. Early in the

1900s what is now known as the USDA was called the U.S. Department of Agriculture and Rural Development. By the 1920s the British colonial service had a field handbook on rural development outlining most of the steps followed in current models. In the 1940s the development of rural Taiwan was begun under the control of the Joint Commission on Rural Reconstruction. In India in the 1950s many of the same processes were fostered under the guise of the Village Development Program supported by various aid agencies. Each of these programs had its own special characteristics and biases. Some were more authoritarian, others relatively laissez-faire, some strongly oriented toward equity and social development, others more concerned with efficiency and commercial growth. The underlying assumptions or models were, however, essentially the same for all.

This chapter outlines the precepts and practice of rural development as a model and reviews the largest and most ambitious of government-supported programs, those financed by the World Bank in a large number of developing countries, beginning in the late 1960s. It describes some aspects of the twenty years of experience of both the Bank and the governments of its borrowing members with the implementation of this strategy.[1] The chapter then examines the Bank's experience in relief against the general model and highlights the lack of attention to institution building in Bank-supported programs. It evaluates some of the problems encountered by governments and concludes with a summary of implications for government as an instrument in rural development.

Rural Development in Perspective

Economic policies tend to guide government action in the development effort because development is seen in essentially economic terms. Underpinning these policies are various models, which attempt to describe the process by which a traditional, rural society is transformed into a modern one. The goal is to identify appropriate interventions and to devise policies that encompass not only purely economic factors but technical and social ones as well.

[1]In describing this experience, the section draws heavily on the report entitled *Rural Development: World Bank Experience, 1965–86,* prepared by staff of the Operations Evaluation Department of the World Bank in 1987. See World Bank (1988).

The Economic Framework

Despite the magnitude of the problems of rural poverty and the extensive efforts to offset it through rural development programs, the economics of rural development is something of an intellectual orphan. As an economic process, rural development involves, among other things, mobilization of resources, expansion of output, increased efficiency of resource use, technological innovation, and a changed balance between and within sectors. But the economic models that underlie and explain the processes of development focus on the technicalities of this process, such as capital creation, investment, resource transfers, surplus generation, and productivity growth. Measures of successful development tend to be represented in aggregates or gross averages, such as new capital formation, export earnings, and GNP per capita.

Perhaps not surprisingly, therefore, traditionally there has been an emphasis on urban and industrial components of economic change, an approach that persists in general thinking. The role of agriculture has been increasingly recognized but usually (especially in development theory) as simply a particular cog in the aggregate economic effort. In this scenario the function of agriculture is to contribute to the rest of the economy, generating a food surplus, providing capital, and releasing labor; Kuznets (1966), for example, stressed the role of agriculture in terms of its contribution of products, resources, foreign exchange, and markets to the development process. This view entails a focus on developing a more commercial, large-scale "modern". agricultural subsector to deliver these contributions. By contrast, the poor, technically inefficient, static, traditional agriculture subsector tends to be treated as an inherited disability that has to be outgrown. Presumably the assumption is that either traditional agriculture will gradually improve or the people will move away.

This approach to economic policy leaves out the people who have to find a livelihood in rural areas (often most of the population in the developing country), who remain surrounded by poverty and who are subsistence-oriented, with low productivity and significant constraints on adapting to rapid change. The philosophy of rural development focuses specifically on helping these people, on ameliorating poverty, and on maximizing the efficiencies of their small-scale operations. The intention is to capitalize on their capabilities and to make rural change a cause rather than a consequence of development.

Crucial to the design of rural development projects (or in the Bank's terms, "the development of a rural development strategy") is an understanding of why people are poor. In a modern economy, open to international trade, poverty is in economic terms an unnatural state. It corresponds to having a large supply of very cheap labor, which entrepreneurs can convert into tradable goods at a substantial profit. In a real sense, poverty thus needs to be explained.

Frequently, it is not necessary to look beyond macroeconomic policies, particularly overvalued exchange rates, inflation, price controls, and various forms of market restrictions, to explain the persistence, or even emergence, of rural poverty. But other causes of poverty are physical remoteness, high transport costs, and government interference with commercial activity. In many cases private investment is severely hampered (or simply prohibited), leaving only state-operated monopoly organizations to carry out the functions of commercial activity. Even if local trade is permitted (usually because it cannot be controlled), governments often restrict or prohibit interregional trade.

Yet another possibility for explaining the persistence of poverty is an extremely restricted or declining resource base. People may not have access to land, or the land may be of such poor quality that existing low capital technologies cannot provide an income. This situation is typically associated with rapid growth in population and a severe lack of infrastructure, which compounds the remoteness of rural people and constrains possibilities for technical change.

Conceptual Models for Rural Development Projects

Such are the circumstances that rural development projects and programs must be designed to overcome. Various models have been promulgated to deal with the factors that entrench poverty, but when stripped of certain special features, they all have major characteristics in common. Conceptually, such projects can be explained through the use of three submodels: an economic or incentives model, a technical or production model, and a social or institutional model. These models are closely interrelated and mutually supporting, and they vary in practice from one situation to another.

The Economic Model. The economic or incentives model underlies the others. Poor villagers typically have a household system in which resources are stretched and incomes precarious, and they are therefore highly risk-averse. They require high returns from any innovation to offset the risk associated with its adoption and the extra effort often required of family labor. Field trials have revealed that smallholders may often require an increase in yield that will pay a return of 200 percent to ensure adoption of a new seed and fertilizer package. For farmers to make larger investments or join a broader program, the incentives have to be substantial. Rural development programs thus work best in areas where some changed circumstance has arisen that greatly increases the villagers' opportunities to produce and sell a salable surplus or other resources (usually labor). In other words, such programs work best in localities where some positive externality has been created that can be internalized to participant households.

This changed circumstance can take many forms. It may be the building of new infrastructure such as an irrigation canal, which is the case with many projects in India. The increased water available can ensure a crop in the regular growing season and often permits a second crop. With these kinds of incentives in front of them, smallholders can be persuaded to make significant changes or incur high costs, including giving up some of their land for field channels. Similarly, it may be the building of some other infrastructure such as a road, which gives access to markets previously out of reach. A nice example is provided by the building of the Karakoram Highway through the Northern Area of Pakistan, which has transformed (with the help of a well-conceived rural development program) the villages of this backward, desolate, and formerly isolated area.[2]

Alternatively, the opportunity may arise with the emergence of a new market opportunity. The rapid growth of a market for high-quality food (including livestock products and fruit and vegetables) in the Persian Gulf countries created substantial incentives, in particular for livestock producers in western Turkey and irrigated farmers in Pakistan. In other settings, the market opportunity may be related to the building of a processing plant for a high-value crop, as was the case with tea in Kenya, cotton in Côte d'Ivoire, oil palm in northern Malaysia, or temperate fruits and vegetables in Himanchal Pradesh (northern India).

Another way that the necessary incentives may be created, perhaps the most widely discussed, if not relied upon, is through the introduction of improved production technology—either locally new or totally

---

[2]For a description and evaluation of this program, see World Bank (1987a).

novel. The innovation is generally a product of agricultural research, as in the case of the high-yielding wheat and rice varieties in much of Asia. But the technical change may also involve the transfer of technology from some other place. It is only rarely, however, that innovation in production technology alone can generate sufficient benefits to provide the incentive for rural development without timely favorable price movements or parallel investments in infrastructure. Rather, new production technology has a complementary role to play, as explained below.

The Technical Model. The technical or production model provides the mechanism by which the benefits generated in the ways indicated above are captured by villages and individual farm families. It relates to changes that take place within the village or on smallholdings. The model involves processes whereby the existing stock of resources, land, labor, and capital are enhanced or made more productive. It might be described as a process of factor augmentation.

Land is made more productive either by improving its fertility, by means of increased water and nutrients (in the form of fertilizer) or by growing higher-yielding crop varieties, or both. It is here that the new technologies produced by agricultural research have a role. Production may be further enhanced by pesticides and herbicides. Less directly, fertility may be enhanced by soil and moisture conservation measures. These elements in combination are typically employed in a complex improved production system. A similar set of changes may also apply to livestock.

Labor productivity is increased by the use of improved tools and equipment or by the use of machines, often together with the introduction of improved and greater power sources for stationary operations (such as pumping, threshing, and milling) and field operations (such as carting, plowing, and mowing). Labor productivity may also be improved by better training, education, nutrition, and health.

Capital is increased and improved by measures to build, rehabilitate, or maintain physical structures and internal village infrastructures. Often village infrastructure has to be added to in order to link into other infrastructure such as roads or irrigation canals built by the state. Improved access by building a link road, bridge, or local canal network can provide a powerful incentive to get a rural development initiative started at the village level. This mechanism has been successfully used and refined in the Northern Areas of Pakistan by the Aga Khan Rural Support Program. Such village works involve the raising of funds, savings, investment, and, frequently, collective work contributions.

The changes in technology or production are to provide the means of increasing output and productivity per hectare and per working day. The associated growth in output may not be large at first, but if production is near subsistence at the start of the process, the growth in marketable surplus can be significantly larger. Often even these increases may be absorbed locally through increased consumption. But ultimately these production gains can be obtained and disposed of profitably only if there are concurrent institutional changes, that is, the growth of a marketing system for inputs and outputs.

The Social Model. The social or institutional model describes the adjustments in village organization, the growth of linkages, and the institutional developments that are necessary concomitants of the changes described in the economic and technical models. In most situations these changes can be divided into two categories: changes internal to the village and changes in the external system, especially government instruments.

Typically, rural areas in developing countries have weak local and government institutions. The feudal fiefdoms have been abolished, the colonial administrations have gone away, and local government has not been established or is weak at best. There are many exceptions, of course; there might be some remnants of feudalism, some highly paternalistic government systems, and some economic and social dualism, with large estates and smallholders side by side. In most of South Asia and Africa these instances are rare, though not unknown. Generally, organization at the village level is weak.

Rural development requires increased organization at the village level. This effort can take the form of village organizations, farmers associations, or smaller subunits. Such local organization is needed to permit collective decisions, to provide a vehicle for leadership, to deal corporately with outside agents from the public and private sectors, to allow the collective construction, repair, and maintenance of shared infrastructure, to facilitate the provision of public services such as education, health, and agricultural programs, and to permit the institution and enforcement of rules regarding such things as water sharing, tree cutting, and payment for shares of maintenance.

The creation of village-level organizations (of whatever form) may require the intervention of a change agent or "animateur" who can persuade villagers to participate, but successful ventures at the individual village level quickly lead to imitation. Achieving full participation often requires concerted effort and incentives in the form of quick results. Since the generation of substantial benefits from the development process usually takes several years, it is often advantageous for governments to provide grants to fledgling local organizations to undertake some significant shared enterprise or remove some constraint affecting most of the community.

Local organizations, once created, first take on traditional functions, but quickly they face new tests in the form of dealing with government (or project) staff and soon after with commercial operations. There are many "tricks of the trade" in getting village organizations formed and able to function reasonably democratically. Such organizations may be involved in a variety of activities, and in different circumstances they may range from being vehicles for local government and selected communal activities to being instruments for collective action, especially in regard to services such as credit and the marketing of produce.

Although outlined here as a logical and orderly process, rural development is typically not so. Not infrequently its proponents are disappointed by the drift in direction they perceive as the supposedly orderly and equitable process evolves. But this outcome is the inevitable consequence of individuals and communities exploiting the opportunities that emerge. Social history tells us that periods of rapid economic change can be rather messy, as was the experience of Georgian England, for instance. The most difficult to achieve of the various goals often pursued in rural development is that of equity in the sense of continued relative economic equality.

Dissatisfaction is often also expressed with the "short-term view" and "low level of advancement," which to some seem to be the goal of rural development. This criticism is largely the result of a misconception. Rural development cannot be perceived as an approach that provides a long-term or permanent solution to development. Economic history tells us that farms must grow in size, landholdings amalgamated, and the population involved in a period of farming decline. One purpose of rural development might reasonably be to provide a stable period whereby younger generations can be educated and find employment beyond the farm and often the village. Not to recognize this important goal may severely restrict the participants in rural development programs.

The New Approach to Rural Development

The 1960s saw a significant shift in development thinking. The persistence of poverty and famine, which was highlighted by the food shortages in South Asia, caused widespread dismay. Disenchantment with the trickle-down theory, which had emphasized a lead role for industrial development, had set in. Agriculture was recognized as a source of food and employment, a source of demand and resources for other sectors, the primary source of revenue in many of the least-developed countries, and even as a primary engine of growth.

Thinking about agricultural development went through a comparable change at this time. First, there was growing acceptance in the 1960s of the potential for expanding farm output using improved technology based on high-yielding varieties of cereal crops that were developed at national and international research centers. The benefits from the new high-yielding varieties were allegedly spectacular and apparently scale-neutral; they were widely applicable (it was only later that the geographic specificity and palatability problems of such technology became fully apparent). There was, however, strong disagreement as to whether this technology and other modern innovations could be used effectively on small farms and whether the direct transfer of Western farm technology favored by many was possible. There was also recognition that the resources necessary for increased agricultural production were predominantly the land and labor of smallholders in developing countries and that within its environment traditional agriculture has comparatively few inefficiencies. The rural development approach had the added attraction that it could solve the problem of food supplies for rural people by helping them to produce their own food. It was recognized, though, that traditional agriculture was severely constrained by a lack of technology, a shortage of improved skills and knowledge, and institutional barriers, such as land tenure problems, which limited access to credit. (With hindsight it is now clear that the effect on production incentives of low administered prices, overvalued exchange rates, and imperfect markets were not well enough recognized at the time.)

Underlying these sectoral debates were the broader issues of equity and income distribution. The rural development strategy did not have redistribution of wealth as a primary objective, but it sought to ensure that a greater proportion of incremental benefits from the Bank's investments would accrue to the rural poor.[3] Land reform was not an integral part of the strategy but was taken up as a separate, parallel policy issue. Further, along with concerns about the pervasiveness of poverty and hunger and concurrent with the growth in

[3]This goal was consistent with the broader philosophy of growth with redistribution accepted by the Bank. See Chenery et al. (1974).

welfare services in Western countries in the 1960s, it was widely advocated that the availability of health, nutrition, education, shelter, and related services were basic needs that should be regarded as social investments to be included in development programs. The implication was that rural development projects should also include provision for such social services as were necessary to ensure that basic needs were met.

All of these factors caused the Bank and other development agencies to reconsider their policies during the late 1960s and early 1970s. Initially, the Bank was strongly influenced by the activities and strategies of other agencies involved in rural development; in due course, the Bank influenced them.

Broadening the Strategy

In the early 1970s, the World Bank changed its approach as it expanded the scale and scope of its activities, recruited many new staff (some from those same agencies), and gained in relative importance. The result was a rural developmental strategy that was multidimensional in its rationale, objectives, methods, and components. Its supporters were almost as various as they were numerous; different proponents often favored one dimension over another. Consequently, the strategy tended to be "all things to all people," a broad enabling mandate, rather than a structured, selective, and differentiated set of policies and priorities. As a result, individual projects and even whole programs were easily diverted to meet the objectives of particular players, either in the Bank or in the governments of the borrowing member countries.[4]

In the midst of a multiplicity of ends and means, the concept of integration appeared self-evidently essential and was widely promoted, with the implication that rural development could be pursued effectively only if the various production and consumption-oriented services were introduced concurrently and in an integrated fashion. The apparent synergy to be realized from integrated rural development was seriously challenged with respect to the feasibility of integrated implementation, but when the capacity for local implementation was a constraint, it was argued that expatriate manage-

---

[4]The report prepared by the Operations Evaluation Department, *Rural Development: World Bank Experience, 1965–86,* is strongly critical of the Bank for neglecting internal training and external education efforts in the 1970s. See World Bank (1988).

ment could keep the program on schedule. For those concerned with the implementation of rural development projects, this notion was a serious misjudgment. (The Bank did not formally adopt the term *integrated rural development*, but the area development projects—with their broader goals, multisectoral components, and several agencies involved—took on all of the characteristics of the "integrated" approach.)

Rural development was not the only approach being promoted in the late 1960s. The concept of large-scale farm development appealed to those primarily concerned with the production and distribution of food and raw materials for processing. It had the advantage of being relatively capital-intensive, both inputs and outputs could be closely controlled, results were highly visible, and management was relatively easy. The technology developed in Western countries in the decades after World War II was well proven and reliable, give or take a groundnut scheme or two. It was being applied in part or in full in many developing countries, notably in parts of Latin America and the Middle East, but various components, such as the introduction of large tractors in South Asia, were becoming commonplace in many countries. Evidence was gradually emerging, however, that wherever it was introduced, mechanization and development of large farms displaced tenants, increased rural unemployment, and resulted in greater poverty and increased urban migration.[5] With a threefold increase in its lending proposed for the coming decade, the Bank was thus faced with an urgent need for a more acceptable approach and one that addressed issues of both poverty and food production but could still absorb a three-fold increase in investment. (Some have argued that the Bank might have found it difficult to identify enough viable large-scale schemes to absorb the greatly increased funds available, and many African countries in particular would have seen little of these funds.) It was in this context that the rural development strategy was promulgated by the Bank.

In fact, the Bank got into the game rather late, despite its initial efforts in the late 1960s. But this late start had the benefit that the Bank could learn from the pioneering work undertaken with the help of nongovernment and bilateral agencies. Community development concepts generally and the village development schemes in India in particular offered one approach. The Joint Commission on Rural Reconstruction in Taiwan offered another. The Bank was also aware

[5]For an empirical case study, see Donaldson and McInerney (1975).

that nongovernment organizations were well ahead in the evolution of approaches. Three widely differing approaches were seen in the Comilla experiment in East Pakistan (now Bangladesh) supported by the Ford Foundation, the Puebla program in Mexico funded by the Rockefeller Foundation, and the International Institute for Rural Reconstruction, a freestanding foundation in the Philippines founded by James Yen.

The bilateral agencies of the former colonial governments of Britain and France had provided examples of alternative approaches in the Caisse Centrale Programs in Western African countries, the Swynnerton Plan in Kenya, the Lilongwe Program in Malawi, and the Kenya Tea Development Authority. These experiences were evaluated and used as models for the Bank's strategy. These early initiatives, and many others like them, also provided project approaches that the Bank could expand and replicate in these and in neighboring countries, but the scope for such expansion and replication was soon exhausted by the scale of operations and the volume of funds available. As the Bank expanded its lending for rural development in the mid-1970s, therefore, it had to find and initiate its own projects.

Defining the Strategy

The Bank's strategy for rural development consciously adopted a broad-based approach.

Since RD [rural development] is intended to reduce poverty, it must clearly be designed to increase production and raise productivity. RD recognizes, however, that improved food supplies and nutrition, together with basic services such as health and education, can not only directly improve the physical well-being and quality of life of the rural poor, but can also indirectly enhance their productivity and their ability to contribute to the national economy. It is concerned with the modernization and monetization of rural society, and with its transition from traditional isolation to integration with the national economy. The objectives of RD, thus, extend beyond any particular sector. They encompass improved productivity, increased employment and thus higher incomes for target groups, as well as minimum acceptable levels of food, shelter, education and health. (World Bank, 1975, p. 3.)

Elements. The means of raising productivity were necessarily tentative at that early stage, but the essential elements were thought to be the following: new forms of rural institutions and organizations to promote the inherent potential and productivity of the poor; acceleration in the rate of land and tenancy reform; better access to credit; assured availability of water; intensified agricultural research and expanded extension facilities; and greater access to public services. In formulating project concepts, Bank staff recognized that though the prime focus would be on smallholders, the larger farmers, where they existed, would not be excluded from project benefits.

The Bank's staff also recognized the need for a strong commitment to rural development by national governments and a sound macroeconomic underpinning, including favorable policies on prices, public sector expenditure, taxation, cost recovery, land tenure, regional development priorities, and the development of technology.[6] International development aid as well as government policies needed to be reoriented. The need to reorient development policies to provide a more equitable distribution of the benefits of economic development was beginning to be widely discussed, but practical applications had been limited. Few governments had made serious moves in that direction, and an essential element of the rural development strategy was therefore to involve governments in developing poverty-oriented policies. It was obvious that unless national policies were redirected toward better distribution of the benefits of development, there would be relatively little that the Bank or other aid agencies could do to accomplish such objectives. At the same time, the Bank also wanted the international community to reorient its development efforts to support the emphasis on poverty alleviation.

Objectives. At the outset, the Bank's goal, as expressed in its president's speech, was to increase production on beneficiary small farms so that by 1985 their output would be growing at 5 percent annually (McNamara, 1973). In support of this ambitious target, the Bank was to direct a substantial part of its lending toward alleviation of poverty by identifying particular poverty groups who would be assisted in the programs.[7] Although there was reference to rural works for the landless, the strategy focused on smallholders as the main

---

[6]The overriding importance of an appropriate exchange rate is, however, noticeably absent from papers on the rural development strategy.

[7]By February 1975, when the *Rural Development: Sector Policy Paper* was published, this already ambitious productivity target for 1985 had become even more ambitious: to expand productivity "by at least 5 percent per year during 1975–79." See World Bank (1975).

target because they had the resources to increase production. Specifically, the Bank aimed in the period 1975–79 to double annual lending for rural development over the 1974 level by earmarking 50 percent of its greatly expanded agricultural lending for rural development projects.

To keep its rural development lending focused on the poor, the Bank introduced a monitoring system that from fiscal year 1974 classified all agricultural projects either as poverty-oriented (that is, rural development projects) or as having no specific poverty orientation. For monitoring purposes, rural development projects were defined as those in which at least 50 percent of the direct benefits were expected to accrue to the rural target group—those beneficiaries below the poverty income threshold calculated for that country.[8] This classification does not mean that none of the latter projects would benefit any of the poor; in fact, many did so quite significantly.

Project Instruments. From the outset it was accepted that the Bank's new strategy would entail taking more risks than did "standard" Bank operations. Moreover, there was uncertainty on how to achieve the aims. "Neither we in the Bank, nor anyone else, have a very clear answer on how to bring the improved technology and other inputs to over 100 million small farmers—especially those in dryland areas. Nor can we be precise about the costs. But we understand enough to get started. Admittedly, we will have to take some risks. We will have to improvise and experiment. And if some of the experiments fail, we will have to learn from them and start anew" (McNamara, 1973, p. 7).

The rural development approach of the Bank was to be implemented mainly through projects. The *Rural Development: Sector Policy Paper* envisaged four major project activities: minimum package programs, coordinated national programs, area development schemes, and sectoral and special programs.

Minimum package programs were intended to provide modest but broad-based improvements in incomes through increases in agricultural output, with institutional development an important prerequisite to ensure mass participation. In addition to the input package, projects would finance research, extension, seed multiplication, and

[8] A significant amount of work related to rural development and the parallel basic needs thrust went into the definition of a poverty line and the nature of poverty. This effort gave birth to a subfield of nutrition and gave significant stimulus to consumption economics and applied econometrics. See Timmer (1987).

fertilizer distribution systems. Models for this approach included the Minimum Package Program in Ethiopia and the crop improvement program in South Korea.

The minimum package approach, particularly in Ethiopia, stemmed from that country's experience with the comprehensive Chilalo and Wolaita projects. These projects were judged to be too comprehensive and too expensive to be replicated, but the projects provided the experience from which the most viable elements could be extracted to form minimum packages. In many respects, the precursors of the minimum package program were the area development projects referred to below.

Coordinated national programs, based on careful definitions of resources and needs of the target population, were to be directed toward a large spectrum of the rural population. The objectives of these programs included the provision of resources and services to specific areas to increase employment, improvement of social infrastructure, and provision of production services. Regions were selected for participation on the basis of their potential for expanding production and low levels of employment and income. This approach was evident in the programs of Korea, Taiwan, Pakistan, and Mexico in the 1960s. These programs were characterized by careful planning and implementation, phasing of components, and extensive changes in institutions, with an emphasis on farmer associations.

Area development schemes were similar to the two programs described above but more appropriate to areas with complex target groups or unique physical conditions requiring a large number of inputs to be brought together, such as improved seeds, livestock breeds, irrigation facilities, fertilizers and chemicals, credit, storage, transport and marketing services, and pricing arrangements. High-quality staff and management were essential prerequisites. It was envisaged that these projects might often be single-product projects, similar to those for tea in Kenya, cotton in Mali, or coffee in Papua New Guinea. Broader approaches that might be followed were exemplified by the Lilongwe Land Development Program in Malawi, the Comilla Project in East Pakistan, and the Puebla Project in Mexico.

Sectoral and special programs were seen as being tailored to meet specific needs of the rural poor as well as others. Public services were regarded as grossly inadequate in rural areas. It was felt that the income of small farmers could be improved substantially if they were supported by better physical and social infrastructure, particularly rural public works, education and training, and other programs such

Table 6.1. World Bank lending for agriculture and rural development (RD) projects, 1965–86

Fiscal year	Number of agriculture projects	Loans for agriculture (billions of U.S. dollars)	Number of RD projects	Share of RD projects in total (percent)	Loans for RD projects (billions of U.S. dollars)	Share of RD loans in total for agriculture (percent)
1965–73	219	3.2	76	35	0.74	23
1974–79	446	12.5	251	56	6.50	52
1980–86	497	26.0	247	50	12.60	49

Source: Data from World Bank, Rural Development: World Bank Experience, 1965–86, Operations Evaluation Report (Washington, D.C., 1988).

as feeder roads, village electrification, water supplies, health facilities, and small industries. A feature of these programs was that they would provide some of the conditions needed for self-sustaining increases in productivity and income and were complementary to, or even components of, programs with broader objectives. Rural education programs were expected to be functional in serving specific target groups and meeting identified needs, thus pointing toward vocational training rather than basic education approaches. The promotion of rural industry in the context of rural development was seen to merit special attention, partly as a means to provide employment for the landless.

Summing Up the Experience of Rural Development

An assessment of the World Bank's twenty years of experience with rural development begins with a look at simple numbers—the amount and composition of the loans, the designated recipients, by region and sector, the economic rate of return on the loans, the failure rate of projects, and estimates of numbers of people reached by projects and the costs of reaching them. In such an exercise, it quickly becomes apparent that the numbers are not so simple. For example, in the 1974 reclassification of Bank projects, what might have been an agricultural project before became a rural development project, so strict comparisons of performance between the two types of projects are subject to many pitfalls. Suitably cautioned, we can take a look at the performance of the Bank's efforts in rural development and examine the factors that significantly affected the success or failure of a project. Much was learned about the importance of agricultural technology, research, institution building, administered agricultural prices, marketing systems, and the government's level of commitment to alleviate rural poverty.

Profile of Lending

The formalization of the new rural development strategy occurred at a time of substantial increases in all Bank lending operations, especially to the agricultural sector. Aggregate Bank lending for agriculture during the period 1965–86 can be disaggregated into three distinct phases (see Table 6.1). There was approximately a fourfold increase in lending for agriculture in the period of fiscal 1974–79 over the fiscal 1965–73 level. The lending for agriculture doubled again in fiscal 1980–86 to $26 billion, of which nearly half was for rural development. This adds up to a total program of over $50 billion of Bank lending for agriculture over twenty years, supporting 1,162 projects. Of this amount 50 percent was lent for rural development projects covering slightly more than half of the 1,162 projects.

The sectoral composition of lending within agriculture remained fairly constant over the period, with irrigation and area development projects constituting 50 percent of all lending for agriculture and 70 percent of lending for rural development. Area development became the most popular vehicle for rural development, with 86 percent of such projects classified as rural development, making up 40.5 percent of rural development lending for the period 1974–86. The regional pattern of lending for agriculture, however, shows substantial differences in the sectoral breakdown of rural development lending. Area development projects contributed the largest share of rural development projects in Latin America (63 percent of the region's rural development projects), western Africa (62 percent), and eastern and southern Africa (54 percent), while irrigation projects contributed the largest share in the Europe, the Middle East, and North Africa Region (EMENA) (47 percent), South Asia (45 percent), and East Asia and Pacific (38 percent). Half of all area development projects were in Sub-Saharan Africa, and 58 percent of all Sub-Saharan Africa rural development projects were area development, compared with 34 percent for all other regions.

The composition of lending by the Bank and the International Development Association (IDA) shows that Bank loans contributed 65 percent and IDA credits contributed 35 percent of total lending for agriculture, while Bank loans contributed 62 percent and IDA credits contributed 38 percent of rural development lending during fiscal 1974–86.

Evaluating the Performance

Projections made in the *Rural Development: Sector Policy Paper* indicated the Bank's lending for agriculture during fiscal 1975–79 to be at approximately $7 billion for projects, with total costs estimated at $15 billion, and assumed that half would go to rural development. It also proposed that the program of the Bank should reach a total rural population of some 100 million, 60 million of whom would be in the poverty target group. When measured by loan and credit approvals, the quantitative funding targets were more than met by the Bank's rural development operations during 1975–79, with 390 projects involving project costs of $29.8 billion and lending commitments of $11.6 billion.

Data on economic rates of return reestimated at project completion for the 192 audited projects show that the weighted average economic rate of return for the group of 112 rural development projects was about 17 percent compared with 20 percent for the 80 non-poverty projects, which is not a significant difference.[9] The appraisal weighted average rate of return for rural development projects was about 27 percent compared with 24 percent for nonpoverty projects, which indicates that the difference between the appraisal and reestimated rate of return for rural developmental projects, at 10 percent, was considerably larger than that for non-poverty projects, at 4 percent. The shortfall in expected performance is thus much larger for rural development projects than for non-poverty projects. Part of this difference is explained by overoptimistic appraisal estimates of economic rates of return for rural development projects. The shortfall is caused largely by delays in implementation. The failure rate for rural development projects was also higher at 37 percent, whereas it was 21 percent for non-poverty projects. In spite of these results, however, rural development projects as a group performed reasonably well when compared with the estimated opportunity cost of capital of 10 percent (although this number is highly arbitrary).

Data on beneficiaries are, of course, only projections based on a number of related assumptions and therefore may not fully reflect eventual project outcome. The limited change in farm incomes at the early ex post evaluation stage, coupled with their general instability, adds to the difficulty of measuring incomes and predicting what these will be at full development of the project. Even if available, data are often weak. The number of beneficiaries should thus be seen as only indicative of a range rather than as point estimates. The total number of beneficiary families expected at appraisal to be reached in rural development projects during the period of fiscal 1974–86 was 20.7 million of which 13.7 million were projected to be in the target group. This latter figure represents about 53,000 families for each of the 390 rural development projects. A review of 83 completed rural development projects for which the number of beneficiaries at full

---

[9] Weights are equal to project costs at completion.

development was available gave a count of 57,000 families. These data, however, were available for only about 80 percent of projects completed by 1986.

The sample of 192 evaluated projects approved during fiscal 1974–79 shows that the costs per beneficiary family at project completion—about $1,100 for all rural development projects—compare favorably with expectations at appraisal of $960. By comparison, costs per beneficiary family were about $1,400 for non-poverty projects at evaluation.

Following the introduction of the rural development strategy, the Bank continued financing large investments in infrastructure in rural development projects. Reviews by the Operations Evaluation Department of these projects confirm their high capital intensity, particularly those in the irrigation subsector and in the EMENA region and in South and East Asia. Overall, expenditures on infrastructure account for nearly half the project costs for all rural development projects. These expenditures were generally not different at completion than anticipated at appraisal, although variations in the share for infrastructure are discernible by region and type of project. The share of costs that infrastructure received in Africa was surprisingly low compared with the need, reflecting in part the very low level of investment in irrigation in Africa.

Learning from Experience: Project Issues

With twenty years of hindsight, it is possible to identify several major factors that significantly affected the outcome of rural development projects. Chief among these was the lack of appropriate technology for achieving improved agricultural performance. The institutional research capacity within many developing countries proved too weak to carry out adaptive research and develop technical innovations that could be used by small farmers. Too little attention was paid to building local institutions that could coordinate the rural development effort. As previously mentioned, there was insufficient awareness of the impact on agriculture of low administered prices, imperfect markets, and an overvalued domestic currency. The focus thus shifts to the important role of government policies and their impact on the agricultural sector.

Adequacy of Technology. One of the main factors explaining the performance of different rural development projects was the adequacy of technology to generate improved productivity and incomes. High-yielding varieties of staple grains, which in conjunction with institutional and related changes in technology associated with fertilizer and irrigation water, were widely successful in much of South and East Asia and various other areas. But the lack of comparable technology for extensive rainfed areas proved limiting, and this factor was only belatedly recognized.

The experience from audited projects suggests that in general there was a pattern of overoptimism and sometimes even plain error with regard to agricultural technology. Only in a few rare cases was there sufficient caution on the technology issue that projects were delayed, phased (with an initial pilot phase), or dropped. A technical package that would raise productivity was an essential component, if for no other reason than that it was required at appraisal to justify the benefits projected for the calculation. Such a package was identified or sometimes assumed for every project, although it was not always clearly defined and frequently had not been tested in the project's environment.

The most common package included new varieties and fertilizers, but even this simple approach proved to be much less successful than appraisals had projected. High-yielding varieties proved more difficult to introduce to small farmers than expected, largely because of the risks involved.[10] Experience suggests that risk aversion by small farmers is usually justified (being based on a reasonable evaluation of the odds); a relatively long buildup period for projects is required for farmers to overcome it as they become familiar with the innovations. Many of the earlier rural development projects provided single-crop technical packages, which farmers were reluctant to adopt because they found them riskier than the more diversified, traditional multi-cropping systems.

If technology was not available, projects contained research or field trial components either to adapt technology from elsewhere or to develop new technology. But adaptation frequently proved less easy and took longer than anticipated; new technology almost always could not be found and tested in time to affect production during a typical five- to eight-year implementation period. In many parts of Africa, where sorghum or millet is the staple food, technical packages acceptable to the local populations have proved especially hard

[10]The considerable literature relating to this phenomenon is conveniently summarized in Feder, Just, and Silberman (1985).

to find.[11] Furthermore, technical packages developed in the 1950s for the Sahel region lost much of their relevance because of a sustained decrease in rainfall.

New technologies often proved applicable only under limited circumstances and were otherwise inappropriate. Technical packages that showed promise experimentally were frequently not adapted to fit farmers' resources and conflicted with land-use practices, as in the Senegal Sine Saloum Project. Agricultural technology to be applied in the Tanzania Kigoma Rural Development Project had not been adequately tested. The proposed package for the Togo Maritime Regional Rural Development Project was not adopted because trials showed it to be ineffective. Soils had not been surveyed in the Somalia Drought Rehabilitation Project, nor had rainfed cropping been tested in the proposed settlement areas. Expensive inputs were not widely accepted in the Kenya Integrated Agricultural Development Project. The recommended levels of fertilizer and insecticides under the Tanzania Geita Cotton Project were based on research results that had not been tested under field conditions, and the recommendations proved inappropriate for a large part of the project area. Under the Yemen Wadi Hadramawt Agricultural Project, farmers did not adopt high-yielding wheat varieties because they preferred their traditional varieties, partly because of the characteristics of the straw for traditional brickmaking.

These experiences and others indicate that the profitable and reliable technologies suitable for diffusion to small farmers were often not available, especially in areas with lower natural potential. As a result of this situation, adoption rates by farmers were lower than envisaged at appraisal, and the impact of projects was correspondingly reduced.

Research. Especially in low-income countries, there are two main sources of concern for the future: technological innovations are not available to support sustained increases in productivity, and the institutional research capacity for developing them in the future is weak. Many national research systems do not generate findings that are relevant to farmers and, given the time lags in research, it is likely that in the early 1990s the situation in many countries will not improve much. This lack of support for adaptive research is perhaps the most pressing long-run issue for the poorest (and smallest) developing countries.

[11]These and similar experiences are usefully detailed in Carr (1989).

In some African countries especially, there is now a growing port-folio of national research projects under way or planned (partly reflecting the failure of many research components in rural develop-ment projects). The long period of research before useful results are achieved, however, indicates the need for a consistent long-term sup-port program. Such an investment can probably be funded only by outside agencies. In contrast, there are countries, including several in Asia, where effective national research systems already exist which are able to respond flexibly to the changing agroeconomic needs of farmers.

Institution Building. Many projects that successfully achieved their physical targets have been criticized in evaluation reports for their negligible impact on institution building. Institutional development has suffered most when reliance on autonomous or semiautonomous project management units have substituted for, rather than strength-ened, line agencies. Because of the pressure generated by tight sched-ules for implementation, these units often had a short-term outlook and did not contribute to the organization of villagers or to the longer-term effort of institutional development. Their advantage is that in institutionally weak environments they can provide substan-tive implementation capability, especially if there is a substantial pro-gram of infrastructure construction, as has often been the case. They are, in fact, a substitute for institution building. Autonomous project management units have not proven very effective, however, in inter-agency coordination when staffs of contributing line agencies felt they had their own programs to implement. Also, in periods of bud-getary crisis, autonomous units have often been the first to suffer cuts. Yet for lack of local institutional capability, particularly in Af-rica, enclave-type projects may have been judged more successful than others, especially at the early stages, given that Bank criteria are weighted in favor of short-term increases in production and physical completion.

Part of the continuing dilemma with these project units (as ana-lyzed in numerous Bank reports) is that they provide an unsound base for continuing project activities after the reduction in donor support. This places the long-run benefits of projects in jeopardy. A partial solution is to avoid special executing units by implementing separate project components in different ministries as though they were separate projects, but with a coordinating unit, located either in a sector ministry, such as the planning agency or the president's office. This approach, however, also has flaws. The coordinating ministry may give the project low priority. Ministries for individual sectors may have a greater interest in projects for which they are fully responsible. Coordination may not be effective, including that among donors financing different components.

There remains yet another concern. Because of the scale and prominence of Bank-supported rural development projects, too many of these projects tended to divert rather than create additional human resources, particularly in the absence of adequate local staff resources and with poorly performing training components. Tackling the problem of constraints on human resources more effectively, particularly in Africa, remains a challenge.

Factors in Project Performance: Policy Issues

The assessment of the impact of the rural development strategy on the overall economic efficiency of the Bank's agricultural sector lending requires that some fundamental questions be addressed. Do small farmers have a potential at least equal to that of the larger private or public enterprises of the borrower country to use properly the resources put at their disposal? Do Bank projects release that potential more effectively than other approaches?

The economic rates of return discussed above suggest that by and large there has been some loss of efficiency associated with the rural development lending strategy. Although the rates of return for rural development projects seem comparable with those obtained in the agricultural sector as a whole and these results in turn are comparable with those in other sectors, their lower rate of success must adversely affect the overall efficiency of lending for rural development. It is arguable, however, that rural development projects are not a substitute for other approaches but are in fact the only way the bulk of these low-income people could have received benefits.

There are also notable individual cases in which the smallholder approach clearly paid off. These projects include irrigation in the Indian Punjab, cotton expansion in West Africa, smallholder tea in Kenya (although the Bank was a latecomer to the program), the Pitsilia Project in Cyprus, and the Corum-Cankiri Project in Turkey. Conversely, Lesotho Basic Agricultural Services Project (BASP) was the classic example of a borrower that promoted government estates and abandoned the strategy of reaching smallholders.

This discussion brings us to a recognition that much depends on the policies pursued by the borrowing country. Of obvious importance to agriculture are the issues of agricultural pricing and marketing. Some governments are more active than others in attempting to influence agricultural prices—whether by bureaucratic tradition, as in the case of marketing boards, or by a concern for the welfare of farmers and consumers. Some governments have made an effort to promote the private marketing sector and invest in marketing infrastructure—for example, roads, communications, and marketplaces. What a government attempts depends ultimately on the resources available to it and its commitment to investing in rural areas and alleviating poverty.

Agricultural Prices. The issue of price distortions in rural development projects has been frequently discussed in audits conducted by the Operations Evaluation Department of the Bank. Of the 192 evaluated agriculture projects reviewed, 57 projects (30 percent) specifically identified issues related to agricultural pricing. The comparable figures for rural development are 112 evaluated projects of which 33 encountered pricing issues. The frequency of pricing issues encountered in agricultural projects and rural development projects is similar. Overall, agricultural pricing issues identified in projects had positive implications for project performance in only 10 percent of the cases and negative implications in 68 percent of the cases. In another 22 percent of the cases the effect of agricultural pricing on project performance was either mixed or unspecified.

Most of the references to pricing in these projects refer to government-administered pricing in borrower countries in "controlled" sectors. But these administered prices had significant impact on prices of certain commodities and on the general price level of the economy, thereby affecting economywide incentive structures.

Low producer prices were by far the dominant issue. Besides creating production disincentives, they sometimes shifted production patterns undesirably and caused people to move out of agriculture. In some cases, input subsidies partly offset low producer prices, but price distortions remained a problem, particularly in depressing incentives to produce. The influence of exchange rates on agricultural prices, however, generally far exceeded the impact of other policy instruments. This was frequently the case in Africa, where overvalued currencies were often the dominant factor discriminating against the projects and agriculture in general.

Marketing Systems. Marketing issues were closely linked with price policy issues. These linkages extend in both directions. Distortions in prices make marketing of farm products all the more difficult. Inadequate marketing arrangements, in turn, reflect on the market price of the produce. Like pricing, marketing components in rural development projects are concerned not only with delivery of inputs and processing and transfer of outputs, but also with the reallocation of displaced factors to other producers.

It was often assumed that existing marketing arrangements would take care of the incremental activities generated by rural development projects. A review of audited projects shows, however, that inadequate marketing severely limited rural development efforts in many countries (World Bank, 1989). The record is also clear that in many cases insufficient attention was given by the Bank to the adequacy of marketing arrangements. Typically, marketing issues are confronted in rural development projects at the later stage of project implementation when production begins or a surplus is generated.

Lack of attention to marketing issues during project implementation caused serious problems in otherwise sound projects. For instance, the Cameroon SEMRY Rice II Project faced substantial marketing-related problems. Partly for reasons of self-sufficiency, the Cameroon government supported substantial rice production in the north, whereas the major market was in the south. This policy resulted in serious marketing problems. As a high-cost product, SEMRY rice was not always competitive with other domestic and imported rice varieties in the principal southern market.

Marketing arrangements in some rural development projects were disrupted by external events that were beyond the control of marketing agencies. In the Malawi Lilongwe Land Development Program Phase III Project, production was far below target, and maize and groundnuts flowed across the borders into Mozambique and Zambia, where prices were higher. Although the Agricultural Development and Marketing Corporation could improve its practices, the audit found that a major part of its ineffectiveness could be attributed to factors outside its control. Such factors included delayed arrival of imported goods, supply controls imposed nationally, disturbances in local transport systems, and ineffective planning by program managers elsewhere in government.

In some cases, marketing problems arose out of institutional arrangements, such as problems related to marketing boards and regulations that provide guidance on marketing issues. The refusal of farmers to accept a project's marketing plans had adverse effects on the outcome of the project. In the Senegal Terres Neuves II Project, for example, marketing plans were not fully accepted by farmers because they did not reflect relative market prices. Maize marketing, an important component of the marketing plan, never functioned in the project area, and most settler farmers were not even aware that the government was offering a producer price for maize.

Experience in rural development projects raises questions of whether improvement in marketing services should be promoted as a component of a production-oriented project or whether a parallel freestanding effort should be undertaken. There are a number of cases in which production projects established appropriate market infrastructure alongside introduction of new production technology. The Kenyan scheme to draw smallholders into tea production is a good example. Here a special institution (the Kenyan Tea Development Authority) was established in part to organize the provision of necessary inputs and to collect and process the output from participating farmers. (It also took on responsibility for extension services and construction of roads in tea-producing areas.) Similar parastatal arrangements were made for the cotton development projects in Francophone West Africa, for certain settlement projects, for many of the smallholder and estate tree-crop projects, and for some irrigation projects. In other cases, however, major marketing problems have required separate projects or initiatives through policy-based lending.

Government Commitment to Alleviate Poverty. Government commitment to the notion of the rural poor as a target group for project interventions lies at the root of the rural developmental strategy. The Bank's rural development approach could succeed only if it was supported by the policies and priorities of the borrowers. Indifference to the plight of the rural poor could be as destructive as outright exploitation of them, because in many cases discriminatory policies and practices had been in place so long as to appear normal and accepted. Fortunately, in many developing countries, including most of Africa, the health and vitality of the economy as a whole have been increasingly identified since the 1970s with the health and vitality of the predominant rural sectors.

The rural poor, though numerous, have little political power, and an altruistic commitment in the upper governing levels of society to alleviate poverty is lacking in many developing countries. There are few incentives to make special efforts to change the balance of economic power even marginally, especially when the powerful have only recently achieved such power and urban interests are much more concentrated and better organized. Despite the absence of strong commitment in many cases, however, a surprising number of governments have been willing to experiment with rural development at the project level. This interest has been indicated through governments' investments in rural development, sometimes in otherwise unsupportive environments.

During the nine years of fiscal 1965–73, 39 countries initiated 76 Bank-funded projects. In the following six fiscal years, 1974–79, a total of 351 rural development projects were initiated by 66 countries, including 32 that had not previously had such projects. For the more recent six-year period, fiscal 1980–85, the number of rural development projects decreased somewhat to 217, but the number of countries, covering the full range of political and social systems, that approved such projects remained at 63, including 7 countries that initiated rural development projects for the first time. The essential point here is that a substantial number of countries have made the commitment to carry out poverty programs in their rural areas. The depth of that commitment is in question in some cases, however, when measured by the number of projects that had problems that were within the government's power to correct—had the right policies been adopted. It should be noted, however, that the commitment of the Bank to a defective design, perhaps without the full agreement of a government, could sometimes explain the lack of government commitment.

The frequency of inadequate government support as an audit issue over the years has been surprisingly high, even allowing that priorities and governments do change. For the five years of fiscal 1979–83, 79 percent of 53 projects suffered from a lack of government support, and this was judged the most important adverse factor in 23 percent of the projects. Part of the reason for this lack of commitment could be that during the 1970s many governments responded to offers of external donor funds that were already earmarked for the fashionable rural development projects and they did not have much freedom to influence the choice of specific projects.

Recent annual reviews of audits by the Operations Evaluation Department have generally shown an improvement in assessed government commitment. In *Annual Review of Project Performance Results for 1986,* strong borrower support was mentioned as a favorable factor for 57 percent, and as the most important factor for 26 percent, of the 35 projects with satisfactory performance—whether poverty-oriented or not (World Bank, 1987b). Insufficient borrower support was mentioned as an unfavorable factor for 43 percent, and as the most important factor for 19 percent, of the 21 projects with unsatisfactory performance evaluated in 1984. These figures show a substantial improvement over earlier ones, and there was no significant difference between rural development and other projects. Although it is simpler to treat "government" as a whole, experience has taught the prudence of carefully considering the interests, objectives, and incentives of individual parts of governments when designing projects.[12]

Government as an Instrument in Rural Development

It is clear, in precept as in practice, that it is not possible to pursue any development strategy, especially one as diverse as rural development, independent of national government and associated domestic institutions. Further, in the case of rural development, positive commitment and usually some involvement of the government in execution of the project are required.

The role of government, however, is *not* to plan, organize, and administer rural development from the center. By nature, rural development is a decentralized approach, in which participation of beneficiaries in project identification, technology selection, decision making, and resource provision is crucial to the motivation and sustained effort necessary for success. Rather, the role of government is to create an enabling environment that encourages smallholders to respond to expanded opportunities as they might arise. This role implies the need for favorable policies toward agriculture and supportive policies for rural development.

The macro-policy framework, including exchange rates, taxation, interest rates, and the sectoral balance of public expenditure priorities, has proven most important for agricultural development. Experience has shown and studies have verified that overvalued exchange rates seriously impede agricultural growth and may even set it back. The experience of many African countries in the 1970s and 1980s

[12]For a discussion of internal government relations in relation to rural development see Tendler (1982).

provides examples. For instance, the effect of the higher costs and diminished incentives for farmers associated with overvalued exchange rates was just as severe in relation to domestic food crops in Nigeria as for export tree crops in Ghana and many other countries.

The adequacy of public investment planning, including planning of overseas borrowing and the allocation of counterpart funds and recurrent cost financing, is similarly important to the success of agricultural development. Evidence abounds that if the rural sector is left out of investment plans, the economy will be constrained to low rates of growth. This investment in the rural sector necessitates a well-developed and established set of ministries able to undertake this planning, together with the political will to support balanced development across sectors and regions.

Microeconomic policy measures, including input- and output-pricing arrangements that reflect long-run opportunity costs, are a second major responsibility of government if agricultural development is to be sustained. The government must ensure that input and output prices do not discriminate against the small producer, that they remain reasonably stable over time, and that they provide sufficient incentives so that new production initiatives can be fostered.

As noted previously, the distortions that exist in agricultural sector prices (of both macro and micro policy origins) often have strongly adverse effects on development. It has been shown that those countries with the highest distortions generally have experienced the slowest rates of economic growth (Agarwala, 1983). The same analysis, however, reveals that it is not always those countries with the least distortions that have the highest rates of growth. In the 1980s, high-growth countries with moderate distortions included Korea, Malaysia, Indonesia, Tunisia, Brazil, and several others. Although there are various explanations for the experience of individual countries, the evidence in aggregate suggests that selective distortions may be a bearable cost in some circumstances. Ensuring a reasonable price policy environment for a development program, especially one involving large numbers of poor smallholders, may not be possible without incurring some investment of this type.

An adequate government regulatory system is a further prerequisite for development. This requirement applies to all sectors, but it is of particular importance for agriculture and for the commercializing process of rural development. Farmers encounter government regulations regarding the functioning of the financial system, land tenure, commerce and trading, and various reporting requirements. Most regulations seem to cause aggravation at times, but the most serious problem relates to their absence. Either their nonexistence or lack of proper enforcement can worsen the development environment for projects.

The lack of a commercial code is a frequent problem. Effective marketing systems employ a system of weights and measures, grades and standards, weighbridge certificates, bills of lading, storage warrants, and, above all, contract enforcement. If these elements do not exist, they may be substituted for by local customs or traditions or by the introduction of an ethnic minority whose culture includes a code of trust, financial intermediation, and information exchange. In modern states this regulatory role is always largely played by government. Yet developing countries often pay little attention to this function. Local conventions that are highly unsuited to modern marketing are often left unformalized. If regulations do exist, they are frequently poorly enforced. This lack of codes and standards often creates serious constraints on commercial development, particularly if long supply lines are involved, as is the case in many rural development project areas.

The combination of macro, micro, and regulatory policies determines the environment that exists for development project initiatives. A satisfactory policy environment is desirable for all development projects. Hence the cliché that you "can't do good projects in a poor policy environment." For rural development projects, however, the large numbers of people involved, the relative poverty, and the general fragility of the early stages of such projects make this environment of particular importance. As the experience with Bank-financed projects has shown, this facilitative environment requires a concerted government commitment to the concept of rural development, as reflected in policies, in a willingness to direct scarce resources, including trained staff, and in provision of appropriate funding and coordination mechanisms.

Going beyond the policies that relate to the agriculture sector as a whole, rural development requires additional kinds of government assistance. The first might be called supporting services, which would include education and health services, agricultural research and extension, and often information and logistics assistance. Each of these is a necessary concomitant to, but not an integral part of, rural development projects. Such services also help to condition the environment in which smallholder families must pursue their livelihood, but unlike the policy measures described above, these efforts have to be tailored to the needs of particular rural areas and populations.

Health services need to focus on the endemic health problems of rural communities. Education programs need to reflect the age, structure, and physical and spatial distribution of the school-age population. Agricultural research has to address the constraints that most severely impede agricultural production. Extension services must be responsive to the particular knowledge gaps and concerns of rural people. These services require local delivery systems and regional programs. They also necessitate trained professionals. Such programs tend to follow a national pattern and frequently fail to serve the specific needs of rural development programs. Making them work well constitutes a major challenge for governments.

The most basic requirement for rural development projects is for local-level programs to plan, coordinate, and implement the process; and this raises the question of whether governments can *do* rural development. The local-level program must involve the people in the process of determining (or helping to determine) their own future. This involvement requires prolonged consultation and explanation, usually with the practical aim of creating or strengthening local-level organization. It involves training in conflict resolution, collaborative planning, and the setting up of internal decision-making procedures in the many villages of an area. Many different approaches have been tried, but not all are successful. The rewards of such grass-roots development activities are allegedly greater than the frustrations, which are legion. The role of change agent requires great dedication, skill, and above all, patience.

Governments are not always very good at organizing such programs. Many of the most successful have been run by nongovernmental organizations or by parastatal units, which have been able to exercise a high degree of flexibility. In practice, the use of a parastatal body—a freestanding, independently managed government entity—has often been the most successful approach employed in government-sponsored rural development schemes. Such has been the case in many countries located in all regions of the world, for instance the Joint Commission on Rural Reconstruction in Taiwan, the Federal Land Development Authority in Malaysia, the Kenya Tea Development Authority in Kenya, and the CFDT (cotton authority) in Côte d'Ivoire. When programs have been kept within government departments, they have been less successful, generally because of the constraints on their flexibility and the tendency for staff to have to take on regulatory and service functions as well as attempt to act as "animateurs," or agents of change.

Whatever the organizational form of the project entity, the role of local organizations has been critical. These organizations may be composed of local village associations, farmer organizations, brigades, or some other grouping. None seems to be more successful than another, which perhaps reflects the fact that the chosen entity has to fit the local sociocultural environment if it is to work. Leaving such institutions to perform their function without unnecessary interference is difficult for government staff.

The final requirement of government is to provide good information services. This role implies the collection, analysis, and feedback of information within the project area as well as the provision of information from beyond the project—for example, information on stocks, supplies, and prices of commodities produced or used in the area. Such services are typically inadequate in rural development projects, though they would seem to be relatively easy to provide at modest cost.

In summary, the rate of rural development, or whether there is rural development at all, is very dependent on government action.[13] Yet successful rural development presents major challenges for government. Local organizing and institution building is the most challenging. The World Bank's support for rural development, though providing a better outcome than is widely believed, has been notably short on support for institution building. Many lessons have been learned, and the majority of poor people in developing countries continue to pursue their livelihood in rural areas. Much needs to be done in the 1990s if their lot is to be improved.

References

Agarwala, Ramgopal. 1983. *Price Distortions and Growth in Developing Countries.* World Bank Staff Working Paper 575. Washington, D.C.: World Bank.

Carr, Stephen. 1989. *Technology for the Major Foodcrops of Sub-Saharan Africa.* World Bank Technical Paper. Washington, D.C.: World Bank.

Chenery, Hollis, Montek S. Ahluwalia, C. L. G. Bell, John H. Duloy, and Richard Jolly. 1974. *Redistribution with Growth.* London: Oxford University Press for the World Bank and the Institute of Development Studies, University of Sussex.

Donaldson, Graham, and John P. McInerney. 1975. *The Consequences of Farm Tractors in Pakistan.* World Bank Staff Working Paper 210. Washington, D.C.: World Bank.

[13]For an eloquent defense of, and case for, a continued rural development effort, see Lipton (1987).

Feder, Gershon, Richard E. Just, and D. Silberman. 1985. "Adoption of Agricultural Innovations in Developing Countries: A Survey." *Economic Development and Cultural Change* 33, no. 2:255–98.

Kuznets, Simon. 1966. *Modern Economic Growth*. New Haven: Yale University Press.

Lipton, Michael. 1987. "Improving the Impact of Aid for Rural Development." Institute of Development Studies Discussion Paper 223. Institute of Development Studies, University of Sussex, Brighton.

McNamara, Robert S. 1973. "Address to the Board of Governors." Washington, D.C.: World Bank.

Tendler, Judith. 1982. *Rural Projects through Urban Eyes: An Interpretation of the World Bank's New Style Rural Development Projects*. World Bank Staff Working Paper 532. Washington, D.C.: World Bank.

Timmer, C. Peter. 1987. "The Economics of Nutrition." In *The New Palgrave: A Dictionary of Economics*, vol. 3, pp. 688–90. New York: Stockton Press.

World Bank. 1975. *Rural Development: Sector Policy Paper*. Washington, D.C.

——. 1987a. *The Aga Khan Rural Support Program in Pakistan: An Interim Evaluation*. Operations Evaluations Report. Washington, D.C.

——. 1987b. *Project Performance Results for 1986*. Operations Evaluation Report. Washington, D.C.

——. 1988. *Rural Development: World Bank Experience, 1965–86*. Operations Evaluation Report. Washington, D.C.

——. 1989. *Agricultural Marketing: World Bank Experience*. Operations Evaluation Report. Washington, D.C.

7

Rural Development:
Problems and Prospects

Cristina C. David

The 1970s witnessed a major shift in official assistance for agriculture and rural development. The persistence of poverty and malnutrition in most of the developing countries, especially in the rural areas, was attributed to the narrow focus on industrialization in the 1950s and 1960s. Since agriculture remains the predominant source of livelihood in these countries, the view that rapid agricultural growth through its strong employment and growth linkage effects in the entire economy will lead to broadly based economic development gained wide adherence. The Green Revolution in rice and wheat, which clearly demonstrated the potential contribution of technical change to productivity growth, further enhanced the belief that a more balanced sectoral approach to development is not only desirable but feasible. In the midst of the food-grain crisis in the early 1970s, Robert McNamara, then president of the World Bank, announced that the major focus of the Bank's rapid expansion would be on rural development to accelerate the alleviation of poverty, and he strongly enjoined other development agencies to follow the same thrust.

Between 1973 and 1980, flows of development assistance and other official assistance for agriculture more than doubled in real terms (World Bank, 1982). The share of loans for agriculture and rural development made by multilateral institutions rose to nearly 30 percent. Relatively few studies have systematically examined the benefits and costs to developing countries of such project lending. What is known about the performance and the problems associated with agricultural and rural development projects has come largely from bank or donor reviews and has been articulated primarily within and among the donor community (for example, Lele, 1984; World Bank, 1988; Binnendijk, 1989). Chapter 6 by Donaldson, which summarizes the World Bank's approach and experience with rural development, is a useful starting point for discussion.

Lending Targets

Judging from the sectoral shifts and patterns of lending since the late 1960s, the reorientation of the Bank's strategy toward rural development has been largely achieved. The share of agriculture in total lending rose from 5 percent in the 1950s and 1960s to over 25 percent in the 1970s and 1980s as the level of lending in real terms was increasing at 15 percent per year. Of the total lending to the sector, about half is classified as for rural development projects, which is more than the target. Although the rural develpment strategy is designed to increase agricultural production directly and indirectly, the ultimate objective is to reduce poverty. Hence, operationally, rural development projects have been defined as those in which at least 50 percent of the direct benefits were expected to accrue to the rural target group, i.e., those beneficiaries below the poverty income threshold calculated for that country. Other agricultural projects are not specifically oriented toward poverty but may nonetheless benefit the poor.

The core instrument for lending was investment in area or integrated rural development projects. They were intended to increase farm income through higher productivity by bringing together a large number of inputs, such as improved seeds, livestock, breeds, irrigation facilities, fertilizers, and chemicals, and by improving credit, storage, transport, marketing services, and pricing arrangements for a target area. This effort would also be supported by investments in physical and social infrastructure—rural public works, education and training, and other programs such as feeder roads, village electrification, water supplies, health facilities, and small industries—that would be targeted to poverty groups. Table 7.1 presents the distribution of World Bank lending and total project cost by loan instruments for rural development and other agricultural lending from 1974 and 1986. Whereas during the 1965–73 period, agricultural lending was concentrated in irrigation (33 percent), credit (25 percent),

Table 7.1. Distribution of amount of lending and total project cost of World Bank–approved projects for agriculture and rural development, 1974–86 (billion U.S. dollars)

Instrument	Rural development		Other agriculture		Total	
	World Bank	Total	World Bank	Total	World Bank	Total
Area development	6.19	14.45	1.65	4.34	7.84	18.79
Irrigation and drainage	7.54	16.70	3.84	9.93	11.38	26.63
Agricultural credit	1.73	10.97	4.09	13.38	5.82	24.35
Agroindustry	0.26	0.57	2.24	5.81	2.51	6.38
Perennial crops	1.44	3.03	1.00	2.60	2.44	5.63
Research and extension	0.55	1.31	1.19	2.79	1.74	4.10
Forestry	0.14	0.25	1.10	3.10	1.24	3.35
Livestock	0.44	0.94	0.82	2.31	1.26	3.24
Fishery	0.16	0.41	0.18	0.33	0.34	0.73
Agricultural sector loan	0.26	0.61	1.98	5.34	2.24	5.95
Others	0.35	0.75	1.37	4.04	1.91	4.78
TOTAL[a]	19.08	49.99	19.46	53.97	38.54	103.97

Source: Adapted from World Bank, Rural Development: World Bank Experience, 1965–86, Operations Evaluation Report (Washington, D.C., 1988).
[a]Sum of individual items may not be equal to total because of rounding error.

Table 7.2. Estimated economic rate of return (ERR) and rate of project failure of audited World Bank rural development (RD) and other agricultural projects, 1974–79 (in
percent)

	ERR		Failure Rate	
	RD	Other	RD	Other
Region				
East and South Africa	1.6	5.3	75	80
West Africa	11.7	7.4	40	44
East Asia and Pacific	15.2	16.0	29	13
South Asia	30.1	34.5	19	0
EMENA	16.5	14.2	21	11
Latin America and Caribbean	11.9	25.4	23	11
Total	16.7	19.5	37	21
Type of Lending				
Area development	10.4	16.0	51	0
Irrigation	18.4	14.6	10	13
Credit	34.1	28.4	17	9
Livestock	0.8	12.1	60	36
Tree crops	7.8	12.6	43	33
Settlement	14.7	–	29	–

Source: Adapted from World Bank, *Rural Development: World Bank Experience, 1965–
86*, Operations Evaluation Report (Washington, D.C., 1988).

and livestock (13 percent) projects, after 1974, area development
became the second largest type of project (20 percent), second only
to irrigation (30 percent), which continued to be the most important, with credit (14 percent) in third place. Area development and
irrigation projects made up 70 percent of total rural development
projects.

Economic Rate of Return

Although lending targets were indeed met, the issue is whether rural development projects performed reasonably well. The estimated
average economic rate of return (reestimated at project completion)
of selected audited projects—17 percent for rural development
projects and 20 percent for other agricultural projects—is higher
than the estimated opportunity cost of capital of 10 percent. Average
economic rates of return, however, conceal large differences across
regions and types of lending (see Table 7.2).

Not surprisingly, but disconcerting nonetheless, the rates of return for both rural development and other agricultural projects are lowest in the poorest countries. In Africa, these rates of return are mostly below the opportunity cost of capital. Rural development projects in West Africa and Latin America including the Caribbean have rates of return that are just marginally above 10 percent. Moreover, the failure rate, at 37 percent, for rural development projects is quite high. Even the 21 percent failure rate for other agricultural projects is not low. In East and South Africa, 75 to 80 percent of rural and agricultural projects failed; in West Africa, 40 to 44 percent did. In Asia and Latin America, failure rates of rural development projects are also relatively high, ranging from 20 to 30 percent.

The greatest disappointment is in the area development projects, the flagship of the rural development strategy. The estimated economic rate of return (10.4 percent) is only about equal to the cutoff rate of 10 percent, and the failure rate is as high as 50 percent. Rural development projects for livestock and tree crops indicate even lower rates of return and higher failure rates, but these projects have involved relatively small loan amounts.

These economic rates of return are way below estimates of actual rates of return to agricultural research obtained in developing countries. Evenson (1989) reported rates of return of over 80 percent in cereal grains and 50 to 80 percent in other staple crops for international agricultural research centers (those in the Consultative Group on International Agricultural Research (CGIAR) system). For national research, the estimates were lower, ranging from 40 to 50 percent in Latin America and Asia and 20 percent in Africa. Yet lending by the World Bank for agricultural research including extension was only 5 percent of total lending for agriculture. The contribution of the World Bank to the CGIAR from 1974 to 1986 was about $150 million, which is less than 10 percent of its combined lending for research and extension. Rates of return for education are also higher than many of those shown in Table 7.2 and are estimated to be about 28 percent for primary education, 18 percent for secondary education, and 14 percent for higher education (Jimenez, 1986). It is now well established that education has a significant impact on agricultural productivity.

The relatively high estimates of rates of return for agricultural credit should be viewed with caution. The inequitable impact of special credit programs involving a subsidized interest rate is well documented in the literature (Adams, Graham, and von Pischke, 1984).

The high estimated economic rate of return does not take into account the negative effect of subsidized credit on the viability of the formal financial market.

A major part of subsidized agricultural credit has been used to finance farm mechanization in countries where labor is abundant. Mechanization may save labor without necessarily increasing total factor productivity. The adoption of tractors in South Asia, for example, is a well-documented case (Binswanger, 1978). Where mechanization is socially profitable, rapid adoption will take place with little government involvement (Pingali, Bigot, and Binswanger, 1987). Providing subsidized credit for mechanization will only widen income disparities between landless and farm households and between small and large farmers with no gain in efficiency. Gonzales-Vega (1984) and others have shown theoretically and empirically that artificially low interest rates result in credit rationing which favors larger and wealthier farmers, precisely the opposite intention of the intervention. A recent study of formal financial institutions in India also indicated that despite directed policy efforts, all types of rural banks and credit institutions have grown more rapidly in regions with favorable agro-climatic potential and well-developed infrastructure and have avoided unfavorable agricultural production environments with sparse market infrastructure networks (Binswanger and Khandker, 1989).

Although the low interest rate policy permitted by external funds has not substantially affected agricultural productivity, it has had adverse effects on operation of rural financial markets (Adams, 1982). It has induced heavy dependence on the rediscount facilities of central banks rather than bank deposits for the supply of loans, thereby limiting the ability of financial institutions to mobilize rural savings. The low repayment rates in special credit programs have also caused the bankrupcy of otherwise viable financial institutions.

What Went Wrong with Rural Development Projects?

The problems encountered in the implementation of rural development projects have been well articulated by Donaldson in Chapter 6 and in other review documents. These problems can be grouped into those that relate to the domestic policy as well as the international environment and those that relate to project design. It became widely understood only in the late 1970s that the domestic macro- and sector-specific policy structure in many developing countries, as well as the high agricultural protection in developed countries, significantly lowered agricultural incentives (Krueger, 1988). With the collapse of the world commodity markets in the 1980s, the unfavorable international economic environment not only exacerbated the incentive problem but also severely constrained the public sector's ability to provide counterpart funds for the successful implementation of the projects.

One may argue that the above factors are largely unforeseen and beyond what a project mode of lending can address. Faulty project design can, however, be minimized to a significant extent. There were at least three major interrelated problems: lack of appropriate technology, overly complex design, and unsustainability of the intervention. The expectation that there is available technology on the shelf or that the new technology successfully developed for irrigated rice and wheat can easily be replicated in rainfed agriculture proved to be wrong. What reinforced the weaknesses rather than the strengths of the individual components of the project was the attempt to introduce an integrated development approach on a project basis rather than have simpler, single-sector projects integrated into a national strategy framework that would be more compatible with local needs and resource capacity. The unsustainability of the interventions in providing recurrent or maintenance cost and staff capacity for implementation should also have been easily foreseen and taken into consideration in the design of the project.

Why Did the Failures Occur?

At least two major reasons for the failures may be cited. First is the limited understanding of the development process—its economic, political, institutional, and social dimensions. It took nearly three decades to learn and incorporate in the design of financial assistance the lesson that not only the level of capital flows but the efficiency of its use must be addressed. With respect to the latter, there are three key areas of concern: the policy environment that determines the structure of incentives within the economy; the institutional framework that conditions the capacity of the country to increase human capital, produce new technologies, and construct the physical infrastructure; and the international environment that affects the level and stability of domestic economic incentives.

Second is the lack of accountability in either the lending agencies or the borrowing countries. The fact that area development programs are unsustainable because of a lack of qualified personnel, the mismatch of capital and recurrent cost, and lack of political will or because they have been designed to support widespread adoption of technologies that were unavailable or unsuitable stems largely from this problem. Since repayment of official loans is perceived to be guaranteed, the incentive for the Bank's staff is in the end to meet lending targets. With rapidly expanding supply of loans, projects tended to be pushed beyond the effective absorptive capacity of many low-income countries. Unfortunately, it may have been difficult to show high short- and medium-term economic rates of return for investments in education, research, and infrastructure, which strengthens the capacity to grow, but these investments would certainly have had greater long-term payoff than area development projects, for which rates of return can easily be overstated.

On the borrower side, those who are making the decision to borrow (bureaucracies) are different from those who will eventually have to repay (general public). Indeed, the bureaucracy directly benefits in the short run from these bank loans by having a larger budget and, too often, personally through corruption. Because of the political structure of many developing countries and the long-term nature of these official loans, there is no built-in mechanism for accountability that would help ensure socially efficient decision making, and yet the borrowing countries, particularly their future generations, ultimately bear the burden of these costly mistakes.

Although the donor community takes steps to review the performance and identify constraints to higher effectiveness of the financial assistance, the review by the World Bank itself noted:

> Evidence to date, however, leaves a lingering concern that although lessons have been learned, they have not been sufficiently applied to later rural development lending. . . . As a result, because so much of typical rural development experience to date has been negative, the voices of experience—or at least the louder ones—tend to be negative rather than positive. In a banking environment, where lending volumes remain the performance measure, this has tended to encourage discounting the value of experience in favor of drive and commitment. The latter qualities are more conducive to keeping a lending strategy alive and buoyant in the face of misfortunes, but obviously they court serious dangers. . . . Review and restructuring of the Bank's lesson learning and application system along more overt lines would help to improve the quality of op-

erations. It could also help avoid some of the needlessly repetitive failings of the rural development strategy. (World Bank, 1988, p. 55)

The pressure to move projects undoubtedly hinders the improvement of the quality of lending. In another part of that review, it was stated:

> The extent to which RD [rural development] projects became a major part of the resource transfer function of the Bank has been a handicap in dealing with the absorptive capacity issue. Although the pressure on RD lending has apparently eased as the burden of the resource transfer objective has passed to structural adjustment and sector loans, there is still a discernible pressure on agricultural lending volume in many predominantly rural economies. (World Bank, 1988, p. 55)

Conclusion

The shift toward structural adjustment and program lending provides a unique opportunity for fostering policy and institutional reforms in developing countries that will reduce the bias against the agricultural sector and improve efficiency in the provision of support services to the sector. Agriculture in developing countries is penalized, however, not only by domestic policy distortions but also by high levels of agricultural protection in developed countries which lower and exacerbate instability in world commodity prices and trade. For the agricultural sector of borrowing countries, benefits from import liberalization, the central focus of the policy reforms, would have been even greater if these measures had been implemented within the context of reciprocal trade barrier reductions vis-à-vis the agricultural sector of the developed countries (Krueger, 1986).

Although structural adjustment and program lending appear to be administratively simpler than project lending, they are actually no less complex in design and manner of implementation. Indeed, these loans require a higher order of analytical capability and political and management skills within the borrowing countries. Technical assistance for the program design must be provided early in the process so that in-depth studies and design of policy reform can be undertaken, or at least actively participated in, by the nationals. Providing opportunities for policy and institutional reforms to originate from within would result both in better design of policy and institutional reforms and in smoother implementation. Too often some program conditionalities may either be inappropriate or, if appropriate, lack local constituency. Unfortunately, the main beneficiaries of policy reforms, the agricultural sector and its government representatives, are often not on the side of these reforms—at least not actively—simply due to lack of knowledge. It is also unfortunate that import liberalization was implemented in the 1980s, when world commodity markets were depressed and the benefits to the agricultural sector could not be clearly perceived.

Project lending on a concessional basis will continue to have an important role in official development assistance. The experience of the rural development strategy, however, clearly suggests that these efforts must be confined to simpler, single-sector projects that build physical, human, and institutional capacity for growth in the long term. This means increased investment in education, research and extension, physical infrastructure, and other areas. Equity objectives should be a criterion in the allocation of those public investments, but it should also be realized that growth is not necessarily inconsistent with the alleviation of poverty.

Internal reviews and mechanisms for improving the quality of lending are grossly inadequate. A system of external review and evaluation of conceptual and operational issues of foreign assistance, with active participation of qualified nationals (not only officials) from borrowing countries, must be established to increase outside pressure for policy and management reforms in the area of official development assistance. Greater investment in strengthening analytical and research capacity within and outside donor agencies, particularly in recipient countries, is clearly called for to narrow the knowledge gap and increase the efficiency of foreign assistance.

References

Adams, Dale W. 1982. "Disrupting Economic Development through Large Amounts of Foreign Aid." ESO Paper 967. Department of Agricultural Economics and Rural Sociology, Ohio State University, Columbus.

Adams, Dale W., Douglas H. Graham, and J. D. von Pischke, eds. 1984. *Undermining Rural Development with Cheap Credit*. Boulder, Colo.: Westview Press.

Binnendijk, Annete. 1989. "Rural Development: Lessons from Experience." Highlights of the Seminar Proceedings, February 1989, sponsored by the World Bank and DAC/OECD, AID Program Evaluation Discussion Paper 25. Washington, D.C.: United States Agency for International Development, January.

Binswanger, Hans P. 1978. *The Economics of Tractors in South Asia.* ADC/ ICRISAT Monograph. New York: Agricultural Development Council.

Binswanger, H. P., and S. R Khandker. 1989. "Determinants and Effects of the Expansion of the Financial System in Rural India." Washington, D.C.: Agriculture Operations, World Bank, February, Typescript.

Evenson, Robert E. 1989. "Human Capital, Technology, Infrastructure, and Productivity Change in Philippine Agriculture: An Asian Comparative Study." Economic Growth Center, Yale University, New Haven. Typescript.

Gonzales-Vega, Claudio. 1984. "Cheap Agricultural Credit: Redistribution in Reverse." In Dale W. Adams, Douglas H. Graham, and J. D. von Pischke, eds., *Undermining Rural Development with Cheap Credit.* Boulder, Colo.: Westview Press.

Jimenez, Emmanuel. 1986. "The Public Subsidization of Education and Health in Developing Countries: A Review of Equity and Efficiency." *Research Observer* 1 (January): 111–29.

Krueger, Anne O. 1986. "Aid in the Development Process." *Research Observer* 1 (January): 57–58.

—— 1988. "The Role of Multilateral Lending Institutions in the Development Process." *Asian Development Review* 7, no. 1:1–20.

Lele, Uma. 1984. *The Design of Rural Development: Lessons from Africa.* Baltimore: Johns Hopkins University Press for the World Bank.

Pingali, Prabhu, Yves Bigot, and Hans P. Binswanger. 1987. *Agricultural Mechanization and the Evolution of Farming Systems in Sub-Saharan Africa.* Baltimore: Johns Hopkins University Press.

World Bank. 1982. *World Development Report, 1982.* New York: Oxford University Press.

—— 1988. *Rural Development: World Bank Experience, 1965–86.* Operations Evaluation Report. Washington, D.C.

8

Food Aid, Development, and Food Security

Edward Clay

A combination of reasons makes it particularly worthwhile at the beginning of the 1990s to look at the relationship between food aid, agricultural development, and food security in developing countries. In the years following the world food crisis of 1972–74, at least partial international agreement was obtained on measures to increase international food security. In the latter part of that decade, several institutional changes were made, particularly at the international level and in bilateral food aid programs. The result of the World Food Conference of 1974 and the Third Food Aid Convention of 1980 was a virtual reconstruction of the regime for food aid.

Circumstances changed dramatically, however, in the decade of the 1980s. The experience of the African food crisis and its aftermath raised questions about the effectiveness of these international arrangements to cope in the future if there is an extremely serious and widespread period of food insecurity. Coping with shortfalls in local food supplies was made more difficult by sharp price variability in world markets for cereals and other foodstuffs and by the barriers to food security imposed on developing countries by the world agricultural trade regime. Developing countries cannot count on the willingness of donor countries to incur the domestic costs necessary to provide meaningful assistance to those countries seeking efficient means of providing reliable food supplies to their population. The combination of firmer cereal markets and budgetary tightening, particularly in the United States, which provides about half the commodities shipped as aid, has underscored the real opportunity costs of food aid as a developmental and humanitarian resource transfer.

The changing trend in the value and costs of food transfers is a topic treated in the first section of this chapter. The problems of assessing the impact of this food aid on the economies, particularly the agricultural sectors, of recipient countries, is a topic of the second section. Whether food aid is an effective instrument in achieving food security is treated in a general sense and then scrutinized in the next section in meeting its ultimate test, the African food crisis, the dominating set of events in food aid policy of the 1980s. The African drought was perceived as a grave threat to the survival of large populations and to sustained economic development for almost the entire continent. Did the nature and timing of the apparently massive donor response significantly ameliorate the effects of that natural disaster on affected peoples and economies? Did that assistance aid or hamper the subsequent recovery from the drought?

How these questions are answered will determine to a large extent the future of food aid, and the last section ventures to speculate. It is likely that there will be a relatively, if not absolutely, smaller level of food aid. Those transfers will be more closely managed and internationally coordinated as the balance shifts between developmental and humanitarian objectives within constraints of the wider agricultural trade regime and development assistance budget. Whether these changes mean that food aid will be even less relevant to the development process in the 1990s than it was in the 1980s will depend to a substantial extent on whether the policy changes brought about by the availability of food aid can stimulate a broader array of economic activities. The food aid itself, as always, will be marginal to these activities.

The Changing Role of Food Aid

If evidence was ever needed that policy problems repeat themselves in cycles, it is amply provided by the debate over the role and effectiveness of food aid. Although food aid in the 1950s was primarily a surplus disposal mechanism for the United States, the failure of the monsoons on the Indian subcontinent in 1966 and 1967 sharply raised the opportunity costs of providing such aid precisely when it was most needed. Surpluses and surplus disposal returned by the early 1970s, only to disappear in the world food crisis in 1973–74.

But surpluses returned in the 1980s and disappeared yet again. Just what is food aid for—surplus disposal, development assistance, or humanitarian relief? Is it a wonderful instrument that meets two or three objectives with just one expenditure? Or does the real purpose of food aid shift as its costs to donors and benefits to recipients shift with the international market?

The historical record suggests a complicated and changing answer, and trends since the early 1970s document a rising commitment to meeting humanitarian needs. But the volumes of food aid provided still reflect opportunity costs to donors as much as needs of recipients, and the changing composition of commodities—relatively fewer cereals and more dairy products and vegetable oils—reflects these opportunity costs as well. More important, concerns about the effectiveness of whatever quantities of food aid are actually shipped have led to efforts to manage the resource more carefully. Such care sometimes takes the form of closely managed project support rather than more general, and less accountable, program support, and sometimes provision of food resources to private voluntary agencies with more effective grass-roots capabilities. Most surprising, perhaps, is that food aid no longer necessarily originates in a donor country for shipment to the recipient. Innovative programs to purchase supplies in second countries, or even in the country in need, raise serious questions about the boundaries of food aid relative to other forms of assistance. Such questions can only be healthy. Food aid has long suffered from its containment within a self-reinforcing network of food aid officials, recipients, agencies, and analysts. Forcing the question of the effectiveness, even appropriate definitions, of food aid onto more open and general ground will help end this isolation and may well contribute to raising the effectiveness of the resource itself.

The need to reconsider the topic is reflected clearly in several trends shown in Table 8.1 and Figures 8.1, 8.2, and 8.3. Table 8.1 breaks down cereals food aid into three functional categories: project aid, which supports agricultural and rural development, nutrition improvement, food security reserves, and other projects; non-project (or program) food aid; and emergency food aid for famine relief and the like. Although the quantities in each category have steadily increased from the totals in the 1979–80 to 1981–82 period to the 1985–86 to 1987–88 period, the relative shares have changed. Program aid has declined in relative importance by almost exactly the amount that emergency aid has increased, leaving the relative share of project food aid nearly constant (but with 50 percent higher volume in total). The

Table 8.1. Total world cereals aid by category of use, in three time periods

Category	1978–80 to 1981–82 Quantity (thousand metric tons)	1978–80 to 1981–82 Share of total (percent)	1982–83 to 1984–85 Quantity (thousand metric tons)	1982–83 to 1984–85 Share of total (percent)	1985–86 to 1987–88 Quantity (thousand metric tons)	1985–86 to 1987–88 Share of total (percent)
Project aid						
Agriculture and rural development	3,494	13.4	4,001	12.4	5,154	14.0
Nutrition improvement	2,293	8.8	2,643	8.2	3,029	8.2
Food security reserves	226	0.9	794	2.5	128	0.3
Other	303	1.2	616	1.9	713	1.9
Subtotal	6,316	24.3	8,054	24.9	9,024	24.6
Non-project aid	15,900	61.1	18,325	56.7	19,331	52.6
Emergencies	3,800	14.6	5,912	18.3	8,394	22.8
TOTAL	26,016	100.0	32,291	100.0	36,748	100.0

Source: World Food Programme, *Review of Food Aid Policies and Programmes*, annual, various issues (Rome).
Note: The quantity of cereals shipped in each of the time periods is a total for the three years.

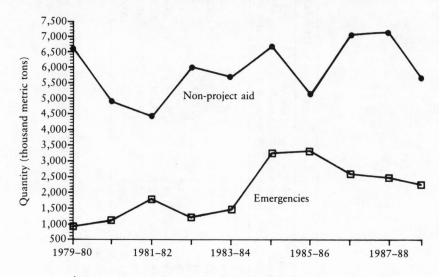

Figure 8.1. Cereals food aid by category of use, 1979–80 to 1988–89

Source: World Food Programme/CFA, *Review of Food Aid Programmes and Policies,* annual, various issues (Rome).

Figure 8.2. Cereals food aid shipments in total and to Sub-Saharan Africa, 1972–73 to 1989–90

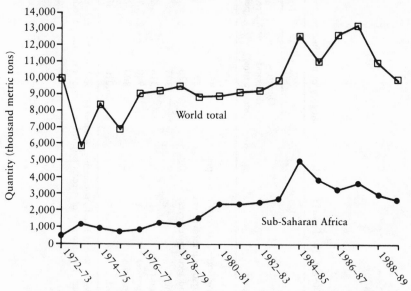

Source: Food and Agriculture Organization of the United Nations (FAO), *Food Aid in Figures,* various issues (Rome).

Figure 8.3. Index of cereals, skimmed milk powder, and vegetable oil food aid shipments, 1977–88

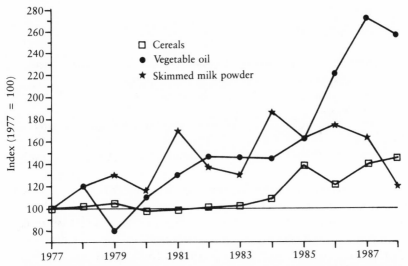

Source: World Food Programme/CFA, *Review of Food Aid Programmes and Policies,* annual, various issues (Rome).

Note: For cereals, the figure for 1977 refers to shipments in the agricultural year 1976–77, and so forth.

reciprocal relationship between emergency food aid and program food aid is clearly evident in Figure 8.1 in which the volumes are shown on an annual basis. Much of the increased emergency food aid was destined for Sub-Saharan Africa, as Figure 8.2 illustrates.

The changing composition of food aid is reflected in Figure 8.3. Volumes of skimmed milk powder rose rapidly in the early 1980s as Europe and then the United States dumped their surpluses in the Third World. But as surpluses were brought under control because of domestic cost pressures, quantities shipped as food aid returned to the lower volumes of the late 1970s. Volumes of cereal aid were nearly constant between 1977 and 1983, reflecting the effectiveness of the Food Aid Convention and relatively stable prices in world markets. But as surpluses and low prices emerged in the mid-1980s, cereal exports as food aid rose by nearly 40 percent. With lower stocks and higher prices at the end of the decade, the volumes declined again (seen tentatively in Figure 8.2). Vegetable oils show the most dramatic growth, as soybean producers lobbied for effective marketing tools to counter the expansion of cheaper palm oil into Third World markets.

What Is Food Aid?

"When I use a word," Humpty Dumpty said in a rather scornful tone, "it means just what I choose it to mean—neither more nor less."

"The question is," said Alice, "whether you can make words mean different things."

"The question is," said Humpty Dumpty, "which is to be master—that's all."

—Lewis Carroll, *Through the Looking Glass,* 1888

In the extensive literature on food aid, the subject is typically taken for granted, and most writers and readers are assumed to share a common definition—that food aid is what the donors choose to call it. Aid-funded food operations in a particular developing country are characterized as food aid. More generally, food aid is what member countries of the Organization for Economic Cooperation and Development report annually to the Development Assistance Committee (DAC) of that organization.

Most writing on food aid is concerned with an aid transfer from a developed to a developing country. The commodity is a foodstuff that a developed country agrees to provide on a concessional basis, which, under DAC reporting definitions, involves concessionality of at least 25 percent. It is an aid transfer because the transaction is mediated by an aid organization—bilateral, international, or a non-governmental organization (NGO)—and therefore is intended to have a developmental or humanitarian objective. In practice, most of the literature on food aid has been concerned only with cereals aid, ignoring the substantial flow of dairy products from the United States and the European Community (EC), edible vegetable oils, fish products, and others. But this common sense definition of food aid has become at least marginally problematic because of a number of developments.

A substantial proportion of what is officially recorded as food aid, approximately 8 to 10 percent of cereals aid in 1986–87 and 1987–88, was acquired by donors in developing countries through purchases with convertible currencies or through a barter or "swap"

Table 8.2. Total triangular purchases or exchanges and local purchases of cereals, 1983–84 to 1987–88

Source	Quantity (thousand metric tons grain equivalent)				
	1983–84	1984–85	1985–86	1986–87	1987–88
Triangular purchases or exchanges	420[a]	706	682[b]	758[c]	771
Local purchase	23	27	140	169	234
Unknown LDC source	0	3	9	28	166
TOTAL	443	736	831	954	1,171
Total cereals food aid (thousand metric tons)	9,850	12,511	10,949	12,579	13,176
Share of purchases in total (percent)	4.5	5.9	7.6	7.6	8.9

Source: Unpublished data from FAO Food Security and Information Service (Rome: Food and Agriculture Organization of the United Nations).

Note: Triangular purchases are purchases by a donor of food commodities in one developing country for use as food aid in another country. Triangular exchanges entail the import of food aid commodities into one developing country where they are exchanged for other commodities which are then exported for use as food aid elsewhere. Local purchases are donor purchases of food commodities in a developing country for use as food aid in that country.

[a]Includes 39,885 metric tons to an unknown destination (so might be a local purchase).
[b]Includes 10,000 metric tons to an unknown destination (so might be a local purchase).
[c]Includes 11,448 metric tons to an unknown destination (so might be a local purchase).

trade arrangements (see Table 8.2). Approximately three-quarters of those commodities were then targeted for Sub-Saharan Africa, and half of the cereals were acquired within this region. Perhaps a quarter of these transactions by volume in Sub-Saharan Africa also involve local purchases in the aid recipient country without any linked trade flow. This partial breaking of the link with trade has led to the suggestion that such transactions are not food aid, but aid for food. Such transactions are usually an alternative to a normal food aid program involving imports by a developing country but which still fulfill donor commitments under the Food Aid Convention.[1] These purchase and barter arrangements raise operational concerns of cost-effectiveness, management efficiency, and appropriateness. In addition, several questions arise with regard to development, as compared with

[1]For an account of the evolution of the Food Aid Convention up to 1980, when a minimum contribution of 7.6 million metric tons was agreed upon by food aid donors, see Parotte (1983). Clay (1985a) reviews the behavior of individual donors in meeting those minimum contributions.

Table 8.3. Cereals food aid shipments in total and to Sub-Saharan Africa, averages for four time periods

Time period	Quantity shipped[a] (thousand metric tons)		Share of total to Sub-Saharan Africa (percent)
	Total	Sub-Saharan Africa	
1972–73 to 1975–76	7,757	831	10.7
1976–77 to 1979–80	9,160	1,229	13.4
1980–81 to 1983–84	9,292	2,524	27.2
1984–85 to 1987–88	12,302	3,971	32.3

Source: Food and Agriculture Organization of the United Nations (FAO), Food Aid in Figures, various issues (Rome).

[a]Figures for quantity of cereals food aid shipped are yearly averages within each time period.

conventional food aid, which concern the "source" economy or region in which the commodities are acquired (Relief and Development Institute, 1987; Clay and Benson, 1990).

Global Trends in Shipments

Food aid levels declined sharply during the early 1970s in the period of the so-called world food crisis. The quantity of cereals provided has never returned to the record levels of the mid-1960s, when approximately 18 million metric tons were shipped, largely supplied under the U.S. PL 480 program (Wallerstein, 1980, Tables 3.2 and 3.6; Huddleston, 1984). Nevertheless, the volume of food aid rapidly recovered to a level of around 9 to 10 million tons of grain or wheat equivalent in the late 1970s (see Table 8.3). In the mid-1980s, the coincidence of the African food crisis and excessively depressed world markets resulted in annual levels of shipments moving slightly higher, to around 11 to 12 million tons in 1984–85. Even higher volumes of 12.5 and 13.6 million tons were shipped in 1986–87 and 1987–88. But with a tighter market situation and the potential for reduced emergency requirements in Africa, total shipments slipped to 11.1 million tons during 1988–89 and 9 to 10 million tons in 1989–90.

The renegotiated Food Aid Convention in 1980—with its emphasis on burden sharing within the donor community and the desire of donors to avoid a recurrence of the negative consequences of a sharp drop in food levels in 1972–74—resulted in relatively stable, institutionally determined levels of food aid (Parotte, 1983; Clay, 1985a).[2] But these levels met a declining proportion of the growing food-import bill of low-income countries. In the mid-1980s the African food crisis coincided with a period of growing surplus disposal problems. Through a combination of supply- and demand-side pressures, levels of food aid began to rise even more. During 1987–88, however, there was a severe upward movement in market prices, first of rice, then of wheat and maize. In 1989–90, food aid has fallen back to levels of the early 1980s.

From the point of view of economic analysis, several problems are raised by this somewhat simplified review of recent trends in food aid. First, it does not take into account the consequences of a major change in U.S. agricultural export policy. During the 1950s and 1960s, the food aid program under PL 480 was the main vehicle for commodity export management and efforts to promote trade with developing countries and some developed countries as well. As commercial exports expanded, first because of market demand in the 1970s and later under the stimulus of various export enhancement programs in the 1980s, United States food aid under the PL 480 program was left as the channel for a decreasing share of agricultural exports. These food aid exports enjoyed significantly greater concessionality but faced increasingly tight administrative restrictions imposed by the Congress.[3]

Second, virtually all food aid since the 1970s, apart from U.S. PL 480 Title I, has been on a grant basis.[4] Consequently, the commodity cost element in the program has involved a higher *planned* element of concessionality. Debts may be rescheduled subsequently, but this is *ex post*, unplanned concessionality.

Third, there has been a geographical reallocation of food aid to least-developed countries, particularly to Sub-Saharan Africa, which has had further implications for costs as well as programs (see

[2]For further discussion of changes in food aid policy and their consequences, see the various papers presented at the Ad Hoc Panel Meeting on "Food Aid Projections for the Decade to the 1990s," October 1988, National Research Council, Washington, D.C.

[3]For example, the share of concessional sales under the Title I Program, mandated to low-income countries as defined in terms of GNP by the World Bank, was set at 75 percent. A fixed minimum level of commodities under the grant Title II Program was established. A minimum mandated share of commodities was to be provided under the de facto grant Title III component of the Title I Program. More recently, minimum targets for monetization of grant food aid were also set. An example of intended increased additionality was the Title III Program initiated in 1979 allowing, in the unhappy language of legislation, "loan forgiveness" when the recipient satisfied policy conditions in the agreement.

[4]The other exception was Japanese aid as so-called Kennedy Round Credits up to the early 1980s provided to Asian countries, including South Korea as well as Bangladesh.

Table 8.3). The concomitant shift in allocations to emergency aid, and to a lesser extent project aid (especially outside Sub-Saharan Africa), also involved donors in meeting both a higher proportion of non-commodity costs and higher per ton noncommodity costs. This is, in part, because of the more favorable terms regarding shipping costs for least-developed countries. There are higher total costs of shipping commodities to landlocked destinations, particularly in Africa.[5] Emergency assistance and project aid are also likely to involve higher administrative costs. The growing volume of acquisition in developing countries has not affected this upward cost trend, since evaluations show that commodity purchases have clustered around the import parity price levels of recipient, often landlocked, countries (Relief and Development Institute, 1987). The overall trend, from the viewpoint of those aid agencies, is for food aid to become more completely a full-cost transfer attributable to aid budgets and for noncommodity costs to grow.

A final complexity for economic analysis in recent trends is the rising importance of non-cereals food aid. Since the mid-1970s, the European Community has been the second most important food aid donor, either in physical quantities or costs as accounted by donor agencies. During a period of twelve years, non-cereals, which were virtually all dairy products, contributed more than half of the accounted costs of EC food aid (Kennes, 1988). The trends in non-cereal food aid are different from those of cereal transfers, and an economic analysis of cereals aid, conceptually and empirically the easiest commodities to study, provides little insight into the impact of these increasingly important non-cereal commodities.

It therefore seems reasonable to conclude of food aid, as with the game of croquet in *Alice in Wonderland,* that things are not exactly what they seem. The cost of food aid transfers, either in financial terms for aid budgets or in real resource costs, may well have been higher during the mid-1980s than in the apparent peak period of the 1960s. A substantial part of food aid since the mid-1970s has not been drawn from surplus stocks of developed countries available for export but has been financed by donors under their commitments for the Food Aid Convention, including, as mentioned above, acquisitions in developing countries. This upward trend in costs of food aid

---

[5]Emergency and project assistance is invariably on a c.i.f. basis. In the case of such assistance to landlocked countries, food is normally provided "rendu destination," that is, up to the border. In low-income countries some donors are also meeting part of the internal shipping and transport costs.

suggests that it would be timely to reexamine the cost-effectiveness of food transfers and also the implications for the scale and balance of commodities to be provided for different purposes.

The trend toward food aid becoming a fully costed aid resource was recognized by the early 1980s. "Food aid is a resource that (at least for the United States) has real opportunity costs. . . . Today when a surplus one year can be followed by relatively tight markets another, food aid cannot be considered a cheap resource. This is particularly true as the United States moves through a period of substantial domestic budget tightening." (Christensen and Hogan, 1982, p. 1). In the mid-1980s, however, the concern about cost-effectiveness subsided, partly because hunger and famine, which temporarily dominated the policy agenda during the African food crisis, became the overriding concern and partly because the links of food aid to surplus disposal had not been entirely broken at the level of individual commodities, as became clear during the period of relatively weaker markets in the mid-1980s.

That food aid was beginning to have real costs was apparent in the case of the disposal of surplus U.S. commodities as grants between 1983 and 1988 under Section 416 of the Agricultural Trade Act of 1949, made possible in 1982 by legislative amendment (Bachrach, 1988). After initial, large-scale disposal of U.S. skimmed milk powder stocks and other commodities as food aid under that provision, allocations were drastically reduced, and during 1987–88 surplus stocks disappeared.

Cereals food aid for the 1988–89 period declined by 18 percent, and further reductions occurred in aid to developing countries in 1989–90 as a result of the tighter cereal market conditions. The major reason is that the PL 480 program is budgeted in financial terms, so the volume of commodities allocated is inversely related to price movements. That negative relationship has implications for food security, which are discussed below.

Economic Development Issues

The crucial test for aid is the extent to which it strengthens the capacity of developing countries to improve humane and material conditions. Assistance to improve the ability to *produce* food would score higher marks than just *providing* food. In some situations, however, the supply of food may be a precondition for sustainable development even in a more long-term perspective. Such a role for food aid has typically been the rationale for "projectized" food aid, such as food-for-work programs, dairy development, and nutrition projects for building human capital.[6]

For more than a decade, the declared objective of food donors has been to make food aid an instrument to promote, rather than hamper, agricultural development. In particular, the intention has been to contribute to increased food security by integrating food aid into a national strategy for food security in recipient countries, particularly in Africa. Measuring programmatic success or failure against such a vast and complicated objective is bound to be controversial and subject to many uncertainties and provisos. Evaluation models used by economists, however, typically require that most variables in the analysis be held constant while only levels of food aid, for example, are permitted to change. More flexible and robust modeling efforts, such as computable general-equilibrium models, require a functional understanding of sectoral interlinkages, which is difficult to acquire even in developed countries and is particularly slippery in developing economies with poor data, imperfect markets, and rapidly changing institutions and economic structures. It is not surprising, then, that these more ambitious modeling efforts have produced few solid conclusions.

Generating Resources for Development

From a macroeconomic perspective, program food aid in kind does not differ, in principle, from other types of commodity aid. It is usually tied to procurement in the donor country, often double and even triple tied, that is, the product and even the producer are identified. Food aid saves foreign currency (improves the balance of payments) for the recipient if the government would have imported anyway. Alternatively, food aid provides an additional supply of food if the government (or private trade) would not have imported the food. Because of the fungibility of such assistance, the criterion for assessing the impact of food aid provided for balance-of-payments support must be the performance of the whole economy.

---

[6]The recent literature on food-for-work is reviewed by Clay (1986); dairy development is discussed by Doornbos, Mitra, and van Stuijvenberg (1988) and by Doornbos, Terhal, and Gertsch (1990). Hay and Clay (1986) discuss the possibilities of using food-for-work for human resource (capital) development. The work by Clay and Singer (1985, chap. 7) includes a review of the literature.

The effects of program food aid are therefore highly dependent on the policy orientation and priorities of the recipient government. If the effect of food aid, in the first place, is to ease the balance-of-payments constraints of the recipient economy, then its effects on development will depend on how additional foreign currency is used. The food is usually sold on the domestic market, as are other forms of commodity aid. The developmental effect of the local currency generated is then determined by how these funds are allocated by the ministry of finance.[7] This uncomfortable reality is fundamental in the aid relationship. The provision of balance-of-payments support and additional budgetary resources are logically alternative, but linked, consequences of food aid. Governments can allocate one or the other of these two resources, not both independently. But the practice of considering these effects sequentially and independently in both impact assessment and policy prescription remains widespread.

Where food aid provides even partial balance-of-payments support, a donor entering into a policy dialogue about the uses of revenue from all local sales is seeking to influence the allocation of the government's *existing* revenues. Most donor governments insist in their declared policies that food aid is provided to satisfy food needs and promote sustainable development, often specifically of the agricultural sector or rural development. But the opportunities for legal or parallel trade open up possibilities for alternative outlets, which dampen the direct effects. Additional cereal imports may promote parallel trade of domestically produced food into a neighboring economy, leaving foreign exchange in the private sector. In addition, government intervention in domestic agricultural markets is almost universal and often pervasive. The potential effects of food aid in freeing foreign exchange may be positive in relaxing constraints on economic growth. Alternatively, these resources may provide a cushion for "bad" nondevelopment policies.

To forestall the diversion of aid from its intended use, donors have tried to exert greater control. Both bilateral and multilateral aid agencies have sought to target aid for the poor, variously defined as specific groups (women, children) or even regions, and to some extent, that explains the continued growth of projectized aid in the 1980s. Program food aid is not particularly appropriate for this

---

[7]This food aid version of the "Parable of Talents" encapsulates the conclusions of many researchers, for example, Isenman and Singer (1977), Clay and Singer (1985), Singer, Wood, and Jennings (1987), and Mellor (1987). The role of food aid in economic development is also discussed in ECC (1981), Stepanek (1979), and Ruttan (1989).

purpose of directly targeting the poorest, or even supporting agricultural development, except via control over government expenditures. The attempt by donors of program food aid to influence the uses made of funds has been apparent in the 1980s, as they sought multi-year programming agreements and specific programmatic uses of counterpart funds to meet local costs of development projects or to facilitate change in economic policy. Program food aid, as with projectized and emergency assistance, thus has been part of the trend for food aid to become a more closely managed resource transfer. There have been an increasing number of multi-year agreements with conditionality in terms of local currencies and policy changes, both on a bilateral basis, as in the early PL 480 Title III agreement in Bangladesh, and the proliferation of multi-donor common counterpart fund arrangements in Sub-Saharan Africa, including Mali, Mauritania, Senegal, and Madagascar (Stepanek, 1979).

On both theoretical and empirical grounds, analysts are skeptical of the capacity of food aid to be an efficient resource for development. Roemer (1989) has restated the economic analysis demonstrating that counterpart funds are not a real resource in any sense; such funds simply allow reallocation within a planned budget or permit inflationary finance for new projects. Empirically, only a relatively small proportion of imports is accounted for by staple cereals, except in a small number of countries in food crisis or in a chronic near-crisis food situation (Sharpley, 1986). Where food aid agreements are contemporaneous with improving sectoral performance (Mali) or general economic performance, food aid has been part of a broader package of financial and commodity assistance, economic conditionality, and domestic economic policy.

"Monetized" food aid is a roundabout way of obtaining the beneficial economic effects of a resource transfer, and a financial transfer would probably have done the trick more easily. The "second-best" nature of food transfers underscores the crucial importance of the additionality question—whether food aid adds to total resources available to poor countries or merely displaces other financial aid—and continuing interest in the cost-effectiveness of food transfers.[8]

---

[8]It is likely that there has been some additionality, probably in the case of the United States. See Stevens (1979), for example. The PL 480 and, more recently, S416 programs have been funded under the Department of Agriculture budget, and the volume of commodities, if not budgeted cost, has fluctuated with availability of surpluses for export. But the Gramm-Rudman-Hollins Amendment has shifted the attributable cost of PL 480 to the foreign assistance budget.

Since the United States has changed its funding process for food aid, it is more likely during the 1990s that food aid will compete at the margin with other elements of the aid program. Both the record of food aid and the options for its use are likely to be more closely scrutinized in the future.

The Economic Impact of Food Aid

The debate on the impact of food aid on the economy, and agriculture in particular, has been inconclusive. It has been difficult to determine the extent to which the food transfer provides foreign-exchange savings or budgetary support. In the former case, there is only a substitution of concessional for commercial imports, usually precluding direct price effects on domestic markets. Methodological problems persist in efforts to assess the direct impact of imports on food production and consumption and the role of other associated agricultural policy measures.

Impact of Cereals Aid. The potential effects of food aid, particularly in the form of cereals, cannot be accurately assessed in isolation, most obviously because of interaction between production and consumption, and also because of the dynamics through time of general economic activity and the evolving forward and backward linkages of the food system (Clay and Singer, 1985, p. 27).[9] Where quantitative economic analysis has been attempted, the overall impact on the economy and the agricultural sector varied considerably among countries, as the impact was heavily determined by the specifics of national food policy: the segmentation of markets, separation of consumer and producer prices, and overall investment policy. Formal quantitative modeling offers the opportunity to explore the role of food imports (supported by food aid) in a wider macroeconomic context.[10] Problems have been identified that justify continuing

[9]For further discussion of the impact of food aid on the economy, see Isenman and Singer (1977), Maxwell (1982), Stevens (1979), Thomas et al. (1989), Cathie (forthcoming), and Sharpley (1986).

[10]Maxwell (forthcoming) and Cathie (forthcoming) use computable general-equilibrium models to indicate a plausible range of outcomes for the impact of food aid to set against the objectives of increased food security and agricultural development. The research effort by Blandford and von Plocki (1977) for India has considerably narrowed the range of probable outcomes. Sectoral modeling suggested that the cumulative effect on agriculture ranged from a possible 15 to 30 percent reduction in food-grain output to an overall increase in agricultural production with higher economic growth rates and food consump-

research, particularly using formal general-equilibrium models or possibly simulation analysis:

> The review of the Indian and Colombian literature . . . underscores the importance of analyzing the macro impact of food aid within a formal general-equilibrium framework. Even in a qualitative sense, the impact remains difficult to establish whilst the analysis is undertaken in one or more of the following ways:
> (a) a partial-equilibrium analysis of the agricultural sector, and especially where the analysis focuses on aggregate food supply:
> or,
> (b) a wider 'informal' general-equilibrium analyis as a series of partial analyses of various potential effects;
> . . . the great weakness of an informal analysis is that, where effects run in contrary directions, the result is necessarily inconclusive. This methodological inadequacy of partial analyses (even where undertaken within a formal general-equilibrium framework) perhaps tempts both a critic and defender of food aid into "straining" the analysis to sustain an unequivocal conclusion (Clay, 1983, pp. 162–63).

The research on food aid has also been limited by its focus on a small number of Asian and Latin American countries that had been long-term recipients of cereals food aid; there is a need to look at the potentially different experience in Sub-Saharan Africa and the other recipients of significant amounts of food aid, such as Bangladesh, Sri Lanka, and Egypt. Analysis of more recent experience, particularly in Sub-Saharan Africa, is under way, and researchers are using either the so-called informal general-equilibrium approach or formal general-equilibrium modeling of African data.[11]

The results of research undertaken so far are broadly consistent with earlier experience. First, the scale of food aid has not been sufficiently large as to have had significant sectoral or macroeconomic impact on many economies. Second, the short-term interaction of environment and public intervention using food aid resources in relatively thin, poorly integrated markets may be spectacular and potentially highly negative. Effects are highly likely to be more localized

---

tion levels. Nevertheless, as Blandford and von Plocki have shown in their careful review of Indian sectoral models, the conclusions to be drawn from such exercises are likely to be highly sensitive to choices in model specification. See Clay (1983, p. 164).

[11]For example, for the "informal" approach, see Maxwell (forthcoming) and for the formal modeling, see Cathie (forthcoming). These authors suggest that both approaches have a place, not least because the limitations of data and the costs of general-equilibrium modeling will limit the number of cases in which that approach can be applied.

than in the larger, more fully integrated food systems of South and Southeast Asia (Maxwell, forthcoming). The policy issue for both government and donors in managing the flow of resources is therefore more difficult in Africa. In particular, the well-understood programming problems of conventional food aid—long lead times, inflexibility in the commodity basket, lack of assurance of continued supply—were again highlighted by experience during and in the aftermath of the African food crisis (Stepanek, 1979).

A further question involving the economic impact of food aid is the overall implications of food imports of developing countries. The greater part of cereal imports is not provided as food aid. A substantial part of these "commercial" or "non-food aid" imports has been financed under soft credit programs of the United States and export subsidization programs of the EC. An assessment of the impact of export programs of developed countries, which would be comparable to the early studies of the impact of the PL 480 program in India up to 1970–71, ought now to take into account programs such as the Guarantee Market Supply (GMS) and Export Enhancement Programs.[12]

Impact of Dairy Food Aid. Food aid in the form of dairy products is, if possible, even more controversial than cereals food aid. Yet these transfers have been relatively neglected in the literature assessing food aid. The findings of the program of research at the Institute of Social Studies, on Operation Flood in India, however, suggest that analytical and policy issues raised are broadly similar to those for cereals aid (Doornbos, Mitra, and van Stuijvenberg, 1988; Doornbos, Terhal, and Gertsch, 1990). The controversy surrounding that program has not ended, partly because protagonists base their views only partially on the record of performance and impact.

A careful sifting of the evidence does show that there has been no dramatic impact on nutrition, either negative in producing communities or positive in consuming urban areas. The additional incomes from milk production have been distributed inequitably, following a pattern similar to the experience of other major Indian rural development programs. There is also great regional variation in performance of institutions and economic impact. Nor can the increases in production that have occurred be attributed more than partially to what has been a very expensive development program. The increase in

---

[12]In 1987, export subsidies on wheat and flour of the United States and the EC countries amounted to about $1.6 billion, or more than half the cost of food aid reported by the OECD member countries, $3 billion.

milk production achieved by the so-called white revolution seems to have been brought about by a combination of factors. Retail price controls have sustained the growth in urban demand, but largely among middle- and upper-income groups. Producers in the marketing, processing, and credit chain have been subsidized. Economists have pointed to the possibilities of using the funds generated through monetization of food aid to achieve an incentive set of prices for producers while not excluding consumer subsidization. But the considerable research and extension effort to increase milk yields through crossbred animals has been a failure (Mergos and Slade, 1987; Alderman, Mergos, and Slade, 1987). Unlike the Green Revolution, technical change has played a minimal role in the growth process. None of the evaluations of Operation Flood undertaken from a donor perspective appears to have been able to conclude unequivocally that the benefits have justified the considerable transfer of resources financed by EC taxpayers through the Common Agricultural Policy and the European Aid Programme.

The Contribution of Food Aid to Food Security

Food security for a country requires both reliable supplies and effective demand. A task for a country's long-term development effort is to create the reliable purchasing power or capacity to produce that generates effective demand for food by all households. The review above suggests that food aid has a marginal role, at best, in this effort. But because food security requires food supplies, many donors, recipient countries, and analysts have argued that food aid could play a much more effective role in helping countries implement their shorter-term policies designed to stabilize food supplies and prices.

The Inverse Relationship: Price and Availability. In general, allocations of food aid by donors have been closely related to their minimum contribution levels under the Food Aid Convention. Nevertheless, because of the behavior of a few large donors, overall levels of food aid have remained sensitive to price movements and relative tightness of markets for most commodities. As a result of the practice in the United States of budgeting the PL 480 program on a fiscal year basis in financial terms, the actual physical allocations and shipments have been sensitive to short-term price movements—downward in fiscal 1979, upward in the mid-1980s, and downward again in 1989 and 1990.

The $416 shipments are even more closely related to availability of surpluses for export. The supply of individual commodities, such as rice or vegetable oil, under the PL 480 program has also been determined on a year-by-year basis, depending on the availability of surplus for export. In accord with the Gramm-Rudman-Hollins Amendment, however, there are now other pressures for reduced financial commitments, which have accentuated the consequences of higher cereal prices on total commodity availability (World Food Council, 1989). Such consequences are clearly illustrated by the substantial year-on-year decline in American pledges to the International Emergency Food Reserve during 1989 (World Food Programme, 1989a).

A close analysis of the short-term movements in allocations by some other donors of food aid also indicates that allocations are sensitive to prices and to the level of national stocks available for export. The pattern of Canadian allocations in the 1980s closely parallels that of the United States. When drought reduced Australian wheat production severely, the government began switching to rice, which, because of the equivalence established under the Food Aid Convention, results in a substantial reduction in actual shipped tonnage of cereals (Clay, 1985a).[13] Allocations of dairy aid—although falling as a result of political decisions to reduce the EC program in the mid-1980s—have fallen even more sharply with rising prices and the disappearance of bulk surplus stocks in the United States and EC in 1988 and 1989 (Kennes, 1988).

Food aid is clearly not a countercyclical element in the world food economy, buffering the effects of market movements on recipient countries. Overall, the aggregated consequence of decisions by donors is to leave food aid still significantly procyclical in the level of allocations and shipments. Bearing in mind that the greater part of food aid is now targeted to least-developed or low-income countries, the dependence of food aid supplies on aggregate food availability is a serious weakness of the existing international arrangements and reflects negatively on the policies of some larger donors. In retrospect, it now appears only fortuitous that the African food crisis coincided with a period of overhanging surpluses. The response might have been more limited, and possibly even more tardy, if there had been a tighter market situation than that which prevailed during 1983–86.

[13]Minimum contributions under the Food Aid Convention are established in wheat equivalent. When these were established in 1980, 1 ton of rice was considered equivalent to 3 tons of wheat. That ratio was reduced by agreement to 1:2.4 for 1988–89.

The Limited Role for Food Aid. The international agreements that made food aid an instrument of international food security, however, have not been a complete failure. The commitments under the Food Aid Convention have resulted in levels of cereal food aid fluctuating at around 90 to 120 percent of the target of 10 million tons established by the World Food Conference of 1974. There is still a clearly procyclical pattern of fluctuations, but it is much less severe than that experienced during 1972–76. That success in committing cereals aid, on however modest a scale, ought to be considered in the light of recent developments and possible scenarios for the 1990s.

The potential need for large imports of food grains from time to time is very great in some countries. For example, if the 1987–88 drought had continued in India during 1988–89, the available scale of food aid would have been insufficient to mitigate the drought. India would have had to rely on large-scale commercial or credit purchases in a tighter market during 1988–89. On the basis of past climatological patterns in India characterized by occasional two-year patterns of widespread drought, this large shortfall in cereals production and need for massive imports remain serious possibilities.

If Bangladesh were faced with a severe, unanticipated reduction of food production, that country, already dependent on food aid, would require additional imports that would be beyond levels at which existing food aid commitments could respond in any significant way. The use of food aid to support a complex of interventions in the food system, including the flexible use of Food-for-Work and vulnerable group feeding programs during the drought of 1979–80 and after the floods of 1984, 1987, and 1988, has been a successful example of support for national food security. Nevertheless, if Bangladesh, which received 1.6 million tons of cereals aid during 1988–89 and imported a total of 2.2 million tons of cereals, were confronted with a drought rather than a flood, import requirements could plausibly be pushed up to a level of 3.5 to 4.0 million tons and above. If food aid were able to cover only perhaps 1.5 to 2.0 million tons of the deficit, which is a plausible assumption if overall levels fall to the 9.0 to 9.5 million tons level of the early 1980s, then the residual import requirements would seriously strain that country's capacity to finance purchases on a commercial basis (as well as push up world rice prices). Those levels of commercial imports costing $400 to $500 million would also crowd out other imports necessary to sustain economic growth.

In the light of these examples, two conclusions can be drawn. First, food aid cannot be considered a significant instrument of food security for countries, such as India, that might require import levels that are nonmarginal in relation to world cereals trade. Second, the transitory problems of food security of countries that are in turn nonmarginal in relation to overall levels of food aid, such as Bangladesh, cannot be assured through food aid. Food aid can be a significant instrument for food security at a national level, or regionally, only in countries that are marginal in their potential import requirement in relation to world cereals trade and also to internationally agreed-upon minimum levels of food aid. Although Sub-Saharan Africa has experienced the most severe, short-term transitory problems of food security, the scale of resources required for countries whose need for food imports is nonmarginal means they cannot depend on regular supplies of food aid for a famine relief effort. Emergency food aid (and increased allocation from other categories of food aid as discussed below) is potentially the most effective response.

These experiences and possibilities justify looking yet again at other compensatory financial mechanisms for providing a buffer against the economic effects of variability in domestic food production and of world market prices on food-import bills. The International Monetary Fund's Compensatory Financing Facility provision for covering abnormal costs of cereal imports was a first step in that direction. The growing role of program lending by development banks and bilateral donors is another area of growing flexibility. But the process is still ponderous. Perhaps an understanding of the economic costs for low-income countries of cereal price variability in the late 1980s will provide fresh impetus to international initiatives in this area.[14]

Management Issues: Food Crisis in Africa, 1982–86

The African food crisis of the 1980s has been the most important test to date of the credibility of the food aid system as reconstructed after the world food crisis of 1972–74.[15] A large number of countries moved almost simultaneously into increasing deficit of staple

---

[14]See Timmer (1989) for a discussion of the benefits of stabilizing food prices through domestic policy interventions.

[15]This account of the African food crisis draws heavily on Borton and Clay (1986).

foods. The crisis began with the onset of drought in the Sahel. Subsequently, the drought became more widespread, including many countries of eastern and southern Africa in 1983–84. The severity of the drought was aggravated by the enfeebled condition of many poorly managed economies severely hit by the second hike in oil prices. There was growing recognition in 1982 and thereafter that in many African countries, agriculture was moving toward a crisis point. Nevertheless, the monitoring of events and articulation of a coherent strategy for responding to that crisis, as it came to involve a large number of countries, were both tardy and initially unsatisfactory.

The Response to the Crisis

Part of the difficulty in devising a response to the famine was that a large number of African countries had poorly articulated and badly integrated (and indeed disintegrating) marketing and distribution systems. Compared with traditional recipients of food aid in Asia, these countries had relatively small populations and aggregate food requirements. Because of the insurgency in Mozambique and the portrayal of the crisis internationally as general and widespread throughout the continent, attention was also probably diverted from the graver conditions rapidly developing in a few countries, particularly Ethiopia, Sudan, Mali, and Chad. Early warning systems were ineffective, and the organization of a response at national and international levels was inadequate and delayed in a number of cases. There was inadequate monitoring of agricultural problems and the nature and extent of stress on vulnerable populations in several countries. From a humanist perspective, there was a criminal failure of responsibility on the part of some governments. Among donors, there was excessive risk aversion characterized by cautious incremental responses until, finally, the entire donor community was galvanized into massive crisis management under the impetus of media coverage of famine conditions in Ethiopia.

On the positive side, many countries, such a Botswana, Kenya, and Zimbabwe in eastern and southern Africa, coped extremely effectively with the effects of drought on food supply and entitlements through a combination of commercial and food-aided cereal imports, increased public distribution and relief operations, and management of markets. Some of the partially affected coastal and landlocked Sahelian countries also coped relatively well. To put the international response in perspective, the relief operations were financed, in order

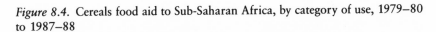

Figure 8.4. Cereals food aid to Sub-Saharan Africa, by category of use, 1979–80 to 1987–88

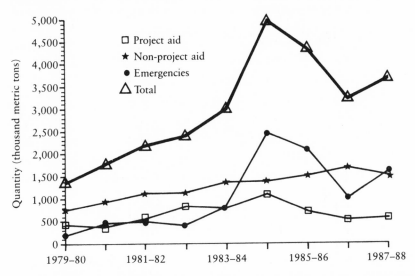

Source: World Food Programme/CFA, *Review of Food Aid Programmes and Policies,* annual, various issues (Rome).

of significance, by the domestic resources of affected countries, aid donors, and last, and very much least, nongovernmental organizations and media-centered fund-raising.

Efforts by Donors. Once the situation was seen as qualitatively different, a continent-wide crisis, there was clear additionality of response (see Figure 8.4 and Table 8.4). During the 1980s, emergency and development programs appear to have been competitive, alternative outlets for cereals food aid within a total that was largely institutionally determined. The only clear exception is 1984–85, when both emergency and program assistance to Africa and the rest of the world increased simultaneously. The lagged response to the crisis, however, resulted in much of the additional emergency food arriving after many lives had already been lost and in some instances when it was no longer required. With favorable rains, production rose nearly to record levels year to year in many of the most severely affected countries, particularly in the Sahel during 1985 and 1986. There is some evidence of possible disincentive effects in Sahelian countries, the Sudan, and, as Maxwell (forthcoming) hints, in Ethiopia.

Table 8.4. Cereals food aid to Sub-Saharan Africa by category of use, in three time periods

Category	1978–80 to 1981–82		1982–83 to 1984–85		1985–86 to 1987–88	
	Quantity (thousand metric tons)	Share of total (percent)	Quantity (thousand metric tons)	Share of total (percent)	Quantity (thousand metric tons)	Share of total (percent)
Project aid						
Agriculture and rural development	655	12.2	1,409	13.5	902	8.0
Nutrition improvement	428	8.0	592	5.7	502	4.5
Food security reserves	184	3.4	461	4.4	111	1.0
Other	102	1.9	310	3.0	316	2.8
Subtotal	1,368	25.5	2,772	26.6	1,825	16.2
Non-project aid	2,849	53.1	3,903	37.5	4,725	42.1
Emergencies	1,146	21.4	3,730	35.8	4,685	41.7
TOTAL	5,363	100.0	10,404	100.0	11,236	100.0

Source: Unpublished data from World Food Programme, Evaluation and Policy Division, Rome.

The massive relief operations were successful in saving lives and limiting distress and economic disintegration in Ethiopia and other affected countries. The scale of food aid operations was such that for many countries, food aid became temporarily a large part of overall development assistance (30 percent of aid from OECD countries to the most affected countries in 1984).

The severity of logistical problems pushed donors and governments closer together in coordinating programming and delivery of food aid through information-sharing and attempts at consistent scheduling of shipments at regional and national levels. At least temporarily, effective local liaison groups involving governments and donors emerged that could have played a valuable continuing role in effective planning of food aid following the crisis. Once the crisis had passed, however, donors and governments appeared to have lapsed back in some cases to more bilateral, less-coordinated relationships.[16] Information-sharing at an international level for food aid to so many countries was unprecedented and encouragingly indicative of agencies coming to terms with more complex multi-donor food aid systems.

Management of the Aid. Nongovernmental organizations (NGOs) provide very little food aid from their own resources. But historically, CARE, and Catholic Relief Services in particular, played an important role administering U.S. project and emergency food aid under PL 480 Title II.[17] During the African food crisis the donor agencies began to use NGOs on an unprecedented scale, both to channel aid and to deliver assistance to affected populations. As a consequence, NGOs appear to have continued to play a relatively large role in organization of food aid to Sub-Saharan Africa, and they accounted for 21 percent of cereals and 30 percent of other food aid channeled to that region in 1988, a higher proportion than that of any other region. NGOs also play unorthodox roles, for example, in cross-border operations (as in Tigre and Eritrea) and provide relief in areas where international and bilateral agencies are not able to operate directly (Southern Sudan). NGOs appear to be playing a larger role in a more closely managed food aid system.

[16]For example, in Tanzania, see Clay and Benson (1988), and also in Kenya (see unpublished interviews by E. J. Clay with staff of donor agencies in Nairobi, January 1986).

[17]Relatively complete data for the World Food Programme's INTERFAIS data base are available only for 1988: from the United States some 15 percent of cereals and 21 percent of noncereals were channeled through NGOs, compared with 12 percent of all cereals and 17 percent of non-cereals food aid (World Food Programme, 1989b).

The Response to the Response

As in the earlier crisis of 1972–74, a traumatic set of events appears to have been internalized as lessons for administrative practice within donor agencies and recipient country governments, as well as NGOs. First, the crisis generated considerable interest in early warning systems as, in effect, a trigger for emergency food aid. Resource flows to such activities have continued at a high level, even if some of the plethora of voluntary initiatives have faded away. Second, food aid donors reexamined their emergency procedures and changed regulations to streamline responses. Third, donors also appear likely to respond more rapidly to appeals for international emergency assistance in order to avert the human distress, as well as the political fallout, of a repetition of the Ethiopian and Sudanese famines of 1984–85. Emergency assistance, at least temporarily, has come to have a relatively higher profile and priority within food aid.

A problematic consequence of that change of priority is that, as noted above, there appears to be substitution between emergency and development food aid, which probably reflects the new priority accorded to the fluctuating scale of requests for provisional emergency aid. If development program food aid has, in effect, become a residual category, that would appear to restrict the potential for a positive effect on development. At best, it would limit the scope for multiyear programming and linking of monetized resources to particular development activities, as envisaged in the 1980s in most proposals for reforming food aid. At worst, it confirms the danger that some bilateral food aid programs serve mostly as all too useful ministerial slush funds for responding publicly to disasters the world over.

The lagged response to the African food crisis has again drawn attention to the cumbersome, slow procedures of many donors. The programming decision process is slow. The mobilization of commodities and the organization of processing and delivery result in total response time from request to delivery, even for so-called emergency assistance, of several months and in many cases more than a year (U.S. General Accounting Office, 1986; Cornelius and Letarte, 1987; Borton, 1989). When food aid is provided in a reactive mode, such inflexibility limits its use in responding to problems of food insecurity and substantial, unanticipated variability in food supply. Recognition of that inflexibility has also stimulated a variety of attempts to make food aid more responsive to varying food situations at regional, national, and international levels.

The use of food aid to build up or finance food security or emergency stocks directly in individual recipient countries had patchy results. Nevertheless, such reserves at a national or regional level, financed by donors, continue to find favor in recipient countries. Donors of food aid and international funding organizations are concerned about issues of food system management as well as costs.[18]

The prepositioning of relatively small stocks in convenient shipment points for rapid emergency responses could obviously reduce the post-decision lead times for emergency actions. Apart from very small World Food Programme stockpiles in Amsterdam and Singapore, little has been attempted, and donors have lost interest in establishing such stocks.[19] "Borrowing" from development programs by rerouting ships is a more practical, lower-cost response. But such an approach sacrifices development objectives to meet immediate overwhelming humanitarian priorities.[20]

Food-for-work or nutritional supplementation projects that can rapidly expand or contract according to the food and economic situation have played an important role in providing a food security net, particularly in South Asia. The most notable example is Bangladesh, where the expanded Food-for-Work and the vulnerable group feeding programs have been used on a vast scale to create employment and supplement income following disaster. The sheer volume of commodities involved also make these operations a factor in management of cereal markets (Clay, 1985b; Hossain, 1987; Relf, 1987). The targeting of additional food after the floods of 1984, 1987, and 1988 was influenced by famine vulnerability maps, as well as the reported incidence of flood damage.

The so-called concertina project concept makes it possible to use an agricultural or human-resource development project in a relief context. The provision of relief *in situ* has positive implications for development, because it prevents socioeconomic disintegration through the displacement of people and disposal of assets and it helps ameliorate the effects of food insecurity. But experience in Asia—for example, Bangladesh—as well as in Africa raises doubts concerning trade-offs

[18]World Food Programme (1984) evaluates earlier disappointing experience with donor financing of emergency food reserves. Several proposals for regional emergency reserves have been made.

[19]For example, proposals for prepositioning were floated at the FAO Committee of Food Security in 1986 but were stalled for lack of donor interest. Apart from concern about stocks, there is continuing skepticism about the feasibility of being able to use prepositioned stocks. Such stocks might have to be drawn down and reexported from a country that might encounter problems in food supply at the same time as other neighboring economies.

[20]When such diversion from developmental food-for-work projects to emergency operations occurred in 1984 in Ethiopia, households in communities dependent on food-for-work found their food security undermined in a period of continuing drought and were obliged to seek relief.

between these gains and the effectiveness of food-for-work activities in particular in creating rural infrastructure and income-generating assets. The quality of labor-intensive investment is likely to suffer. Whatever can be undertaken in a crisis on a larger scale is less well designed and managed than other programs. Development projects may also acquire a relief, make-work character.

As a result of recent experience in Africa, there appears to be an increased awareness of the problem of the appropriateness of commodities and of the need for more general flexibility in the acquisition of food aid commodities. The inappropriateness of distributing wheat and rice in rural economies that are based on coarse grains and tuber crops has stimulated triangular, local purchase and swap operations, which were discussed above. Such flexibility in financing food acquisition from unconventional sources increased considerably in response to the appearance of temporary surpluses in public and private stocks in many African countries in 1985–86. Several donor agencies have modified their procedures to mandate at least a limited program of commodity acquisition in developing countries, and such practices are likely to become more commonplace in food aid operations.

One final doubt must remain concerning the food aid response to the African emergency. The crisis occurred in a large number of countries whose aggregate consumption of staple foods is relatively modest compared with that of the large, densely populated countries of Asia. The food aid response involved only a maximum 20 percent increase in cereals food aid during a period when world markets were overhung by large stocks and when exporters were competing with increasing intensity for markets. The response to that crisis, therefore, provides little indication of how the food aid system would have coped with a larger-scale food crisis in Asia that might plausibly have been linked to a rapid movement in world market prices.

Future Food Aid Policy: A More Closely Managed Resource Transfer

A majority of the voters in rich countries believe that food aid is an effective vehicle for assisting poor countries because it is the one program that seems to resolve the paradox of food surpluses in a world of hungry people. Disillusionment among professional development specialists with the actual record of food aid in feeding hungry people in the short run and making them less vulnerable to natural and political disasters in the long run has not translated into sharply diminished political enthusiasm for using food aid whenever possible. As Eastern European economies crumbled as fast as the Berlin Wall, food aid was the immediate response from the United States and the European Community. Late in 1989, the pork belly market in Chicago rallied sharply when it seemed the United States would ship 20 million pounds of bacon to Poland, and it collapsed just as quickly when bureaucratic delays seemed to threaten the effort.

Governments of donor countries are, however, sensitive to charges that food aid has not benefited the poor or helped the development process. The response typically is to mandate further controls on who may receive food aid, in what quantities, and how it may be used in the recipient country. Food aid negotiations increasingly are conducted as part of a broader policy dialogue, with food aid commitments conditional on changes deemed desirable by the donors. This increased managerial control and monitoring has raised the cost of providing a unit of aid in the form of food, as opposed to simple financial resources, and food aid has become a relatively smaller element of overall development assistance. There would seem to be a strong case for concentrating this limited resource on a small number of practical goals and in the neediest countries. Supporting structural adjustment in the poorest African countries would meet both objectives.

A substantial fraction of food aid resources has already been redirected to Africa, but the reactive nature of this response is reflected in the high proportion of emergency assistance. The difficult question now is whether food aid can contribute to the rehabilitation of severely weakened economies of Sub-Saharan Africa through short-term balance-of-payments support as well as additional food supplies. Experience in Mali in supporting restructuring of the food sector suggests that food aid can play a positive role in ameliorating the human costs of structural adjustment while contributing to the broader economic restructuring.[21] But the success in Mali has not been replicated, and serious doubts exist about the cost-effectiveness of the approach and its political sustainability.

The proposal to use food aid as an element in structural adjustment is not new. With respect to the attempt by Sri Lanka to reform its food policy after 1977, it was argued that food aid could play "a significant role in alleviating the suffering of poor people caught in the dilemma

---

[21]These various roles for food aid in structural adjustment were the focus of a special issue of *Food Policy* 13 (February 1988), edited by D. John Shaw and Hans W. Singer.

imposed by inappropriate macro prices" (Timmer and Guerreiro, 1981, p. 21). Indeed, a careful analysis of why food aid failed to play such a role in Sri Lanka might offer instructive lessons for food aid policy. A feature of the Sri Lankan experience highlights the broader problems of food aid policy: initial success in restructuring may lead to a weakening, rather than to continuity, of support by food aid donors. The initial, apparently successful restructuring of the Sri Lankan food system between 1977 and 1981 resulted in that country's becoming a lower priority for food aid as other recipients, Sub-Saharan Africa in particular, came to dominate the policy agenda.

An alternative goal might be to use food aid to help the most vulnerable countries pay for the rising volume of food imports that has been projected. Inability to finance important imports will constrain the development process, and extensive malnutrition can be addressed only by expanding aggregate food supplies faster than foreign exchange constraints will permit. Mellor (1987) has emphasized a considerable range of complementary opportunities for using increased food aid effectively both to finance the development process and to be used directly in labor-intensive rural investment, even in Africa.

There are severe practical difficulties awaiting these substantially more ambitious direct uses of food aid for development. The recent record in many least-developed countries, particularly in Africa during and after the recent food crisis, has underscored the difficulties of programming aid-financed imports in those economies that have highly variable import requirements. How is development to be financed in years when food imports are not needed? In addition, conventional food aid does not provide appropriate cereals for many economies that are based on coarse grains and tubers.

Increasing the role of food aid in financing the development process implies that food aid is less costly than financial aid. Even if the juggling of agricultural and aid budgets makes that appear to be so for some donors, the economic resource costs for developed countries are now probably not lower, but higher, than financing food imports or supporting rural development directly. If food aid is provided in increasingly flexible ways, with purchases in developing countries and monetization rather than direct distribution, then the distinction between food aid and aid for food will become less important. Conventional doubly-tied food aid would gradually disappear in the face of such flexible arrangements. Then the food aid debate would no longer need to be distinct from the overall debate over the magnitude and effectiveness of development aid in general.

Acknowledgments

I am indebted to John Borton, Hans Singer, and Olav Stokke for comments on an earlier version of the chapter and to participants at the EADI workshop, "The Performance and Economic Impact of Food Aid" at Lysebu, Oslo, 19–21 February 1989, whose papers and ideas have influenced the drafting of this manuscript. Charlotte Benson provided invaluable statistical assistance.

References

Alderman, Harold, George Mergos, and Roger Slade. 1987. *Cooperatives and the Commercialization of Milk Production in India: A Literature Review.* Working papers on Commercialization of Agriculture and Nutrition, no. 2. Washington, D.C.: International Food Policy Research Institute.

Bachrach, P. 1988. "Opportunities and Risks: An Assessment of the Regular 416 Program." Washington, D.C.: Planning Assistance, Inc. Typescript.

Blandford, David, and J. von Plocki. 1977. "Evaluating the Disincentive Effect of PL 480 Food Aid: The Indian Case Reconsidered." Department of Agricultural Economics, Cornell University, Ithaca, N.Y. July. Typescript.

Borton, John. 1989. "UK Food Aid and the African Emergency, 1983–1986." *Food Policy* 14, no. 3:232–40.

Borton, John, and Edward Clay. 1986. "The African Food Crisis of 1982–1986." *Disasters* 10, no. 4:258–72. Revised version published as chapter 9 in Douglas Rimmer, ed., *Rural Transformation in Tropical Africa.* London: Belhaven Press, 1988.

Cathie, John. Forthcoming. "Modelling the Role of Food Imports, Food Aid and Food Security in Africa: A Case Study of Botswana." In Edward J. Clay and Olav Stokke, eds., *Food Aid Reconsidered.* London: F. Cass.

Christensen, Cheryl, and Edward B. Hogan. 1982. "Food Aid as an Instrument of Development: A Seminar Report." In Cheryl Christensen, et al., *The Developmental Effectiveness of Food Aid in Africa,* pp. 1–9. New York: Agricultural Development Council.

Clay, Edward J. 1983. "Food Aid: An Analytic and Policy Review (with Special Reference to Developments and the Literature since 1977)." Institute of Development Studies, University of Sussex, Brighton. July.

———. 1985a. "Review of Food Aid Policy Changes since 1978." WFP Occasional Paper 1. Rome: World Food Programme.

———. 1985b. "The 1974 and 1984 Floods in Bangladesh: From Famine to Food Crisis Management." *Food Policy* 10, no. 3:202–206.

——. 1986. "Rural Public Works and Food-for-Work. A Survey." *World Development* 14, nos. 10–11:1237–52.

Clay, Edward, and Charlotte Benson. 1988. "Evaluation of EC Food Aid in Tanzania." Report for EC Commission. London: Research and Development Institute. Typescript.

——. 1990. "Aid for Food: Acquisition of Commodities in Developing Countries for Food Aid in the 1980s." *Food Policy* 15, no. 1:27–43.

Clay, Edward J., and Hans W. Singer. 1985. "Food Aid and Development: Issues and Evidence." WFP Occasional Paper 3. Rome: World Food Programme.

Cornelius, J., and M. Letarte. 1987. "Timeliness of Canadian Bilateral Emergency Food Aid and Deliveries, 1983–1984 to 1985–1986." Ottawa: Canadian International Development Agency. Typescript.

Doornbos, Martin, M. Mitra, and P. van Stuijvenberg. 1988. "Premises and Impacts of International Dairy Aid: The Politics of Evaluation." *Development and Change* 19:467–504.

Doornbos, Martin, Piet Terhal, and Liliana Gertsch. 1990. *Dairy Aid and Development: Current Trends and Long-Term Implications.* New Delhi: Sage.

European Communities Commission (ECC). 1981. "Food Strategies: A New Form of Cooperation between Europe and the Countries of the Third World." *Europe Information (Development)* DE 40(x374/82). Brussels.

Hay, Roger, and Edward J. Clay. 1986. "Food Aid and the Development of Human Resources." *WFP Occasional Papers* 7. Rome: World Food Programme.

Hossain, Mosharaff. 1987. "From Relief to Development: A Review of Food-for-Work and Vulnerable Groups Development Programmes in Bangladesh." Paper for the Government of Bangladesh/WFP Seminar on Food Aid for Human and Infrastructure Development in Bangladesh, Dhaka, December.

Huddleston, Barbara. 1984. *Closing the Cereals Gap with Trade and Food Aid.* IFPRI Research Report 43. Washington, D.C.: International Food Policy Research Institute.

Isenman, Paul J., and H. W. Singer. 1977. "Food Aid: Disincentive Effects and Their Policy Implications." *Economic Development and Cultural Change* 25 (January): 205–37.

Kennes, W. 1988. "Food Aid and Food Security: Approach and Experience of the European Community." *Tijdschrift voor Sociaal wetenschapperijk onderzoek van der Landbouw* 3, no. 2:138–57.

Maxwell, S. J., ed. 1982. "An Evaluation of the EEC Food Aid Programme." Commissioned Study 4. Institute of Development Studies, University of Sussex, Brighton.

Maxwell, S. J. Forthcoming. "Disincentives to Agricultural Production." In Edward J. Clay and Olav Stokke, eds., *Food Aid Reconsidered.* London: F. Cass.

Mellor, John W. 1987. "Food Aid for Food Security and Economic Development." In Edward J. Clay and D. John Shaw, eds., *Poverty, Development, and Food*, pp. 173–91. London: Macmillan.

Mergos, George, and Roger Slade. 1987. *Dairy Development and Milk Cooperatives: The Effects of a Dairy Project in India.* World Bank Discussion Paper 15. Washington, D.C.: World Bank.

National Research Council. 1988. *Food Aid Projections for the Decade to the 1990s.* Report on an Ad Hoc Panel Meeting, 6–7 October. Board on Science and Technology for International Development, Office of International Affairs, National Academy of Sciences. Washington, D.C.: National Academy Press.

Parotte, J. H. 1983. "The Food Aid Convention: Its History and Scope." *IDS Bulletin* 14, no. 2:10–15.

Relf, C. 1987. "Food Assisted Rural Employment and Development in Bangladesh: Proposals for Institutional Change." Paper for the Government of Bangladesh/WFP Seminar on Food Aid for Human and Infrastructure Development in Bangladesh, Dhaka, December.

Relief and Development Institute. 1987. "A Study of Triangular Transactions and Local Purchases." *WFP Occasional Paper* 11. Rome: World Food Programme.

Roemer, Michael. 1989. "The Macroeconomics of Counterpart Funds Revisited." *World Development* 17 (June):795–807.

Ruttan, Vernon W. 1989. "Food Aid: Surplus Disposal, Strategic Assistance, Development Aid, and Basic Needs." Department of Agricultural and Applied Economics, University of Minnesota, Minneapolis. Draft typescript.

Sharpley, J. 1986. "Australia's Food Aid Policy to Africa." National Centre for Development Studies, Australian National University, Canberra. Typescript.

Singer, Hans W., John Wood, and Tony Jennings. 1987. *Food Aid: The Challenge and the Opportunity.* Oxford: Clarendon Press.

Stepanek, Joseph R. 1979. "Food for Development: A Food Aid Policy." Paper for A/D/C Seminar "Implications of U.S. Food Aid: Title III," Princeton, N.J., 15–16 January.

Stevens, Christopher. 1979. *Food Aid and the Developing World: Four African Case Studies.* New York: St. Martin's Press.

Thomas, M., et al. 1989. "Food Aid to Sub-Saharan Africa: A Review of the Literature." Institute of Development Studies, University of Sussex, Brighton. Typescript.

Timmer, C. Peter. 1989. "Food Price Policy: The Rationale for Government Intervention." *Food Policy* 14 (February): 17–27.

Timmer, C. Peter, and Matthew Guerreiro. 1981. "Food Aid and Development Policy." In G. O. Nelson et al., *Food Aid and Development*, pp. 13–30. New York: Agricultural Development Council.

U.S. General Accounting Office. 1986. "Famine in Africa: Improving Emergency Relief Programs." GAO/USAID-86-25. Washington, D.C.: U.S. Government Printing Office.

Wallerstein, Mitchel B. 1980. *Food for War—Food for Peace: United States Food Aid in a Global Context*. Cambridge: MIT Press.

World Food Council. 1989. "Current World Food Situation." WFC/1989/5. 25 April. Rome.

World Food Programme (WFP). 1984. "Sectoral Evaluation of Food Aid for Price Stabilisation and Food Reserve (Emergency Stock) Projects." WFP/CFA: 18/11 Add 1. Rome.

——. 1989a. *Annual Report of the Executive Director: 1988*. WFP/CFA:27/P/4. 19 April. Rome.

——. 1989b. *Review of Food Aid Policies and Programmes*. WFP/CFA:27/P/ INF/J. 20 April. Rome.

9

Whither Food Aid?
A Comment

Walter P. Falcon

Chapter 8 provides a thoughtful, balanced assessment of food aid as it has evolved in theory and practice over the past thirty-five years. This chapter extends Clay's historical assessment and examines three dimensions of future food aid: the effects of the Uruguay Round of trade negotiations, the impact of new U.S. agricultural legislation, and the prospects for improved implementation of food programs. For reasons of complementarity, these comments also focus mainly on donors, and especially on the role of the United States.

History of U.S. Food Aid

A cynic would summarize the history of U.S. food aid in three phases: "mostly grains," "mostly to Asia," and "mostly to friends." Although this assessment is correct as far as it goes, a current review of food aid would indicate four trends that are also worth noting. First, there has been a regional switch in total food aid to Sub-Saharan Africa. In 1975, for example, only about 10 percent of global food aid went to that region; by contrast, total food aid shipments to Sub-Saharan Africa in 1985 were up by a factor of seven and constituted about 45 percent of global shipments (Schultz, 1987, p. 139). The regional pattern of U.S. shipments changed much less, however, and the 1989 foreign assistance request to Congress showed that less than 15 percent of the combined total food aid under Titles

I, II, and III was destined for Sub-Saharan Africa.[1] These relative changes reflect both the varying colonial ties of the United States as opposed to Europe and the increased role of donor countries— mainly European—in total food aid tonnages.

Second, there has been some progress in allocating food aid to the least-developed countries. U.S. congressional restrictions that mandate two-thirds or more of the food aid be given to countries with per capita GNP of less than about $800 have done much to strengthen the potential link between alleviating hunger and using food aid. Nevertheless, the alleged national security dimensions of U.S. food aid are obvious. Egypt still receives 25 percent of all food aid under Titles I and III, and El Salvador receives more than 5 percent. Both programs underscore the multiple purposes for which food aid is given.

More generally, the absence of strong linkages between countries with large numbers of undernourished people and U.S. shipments of food aid is a source of continuing concern. There is broad agreement among hunger specialists, for example, that about three-fourths of all the hungry people in the world are currently in eleven countries: Bangladesh, Brazil, Cambodia, China, Ethiopia, India, Indonesia, Pakistan, the Philippines, Sudan, and Zaire. The 1989 congressional request, however, shows only about 20 percent of food aid under Titles I and III for these nations combined, and none at all for six of them. The official rationalization as to why Egypt has a nutrition-based cereal need of 231 kilograms per capita, while Ethiopia and India, for example, receive no Title I food aid is a curious bit of political arithmetic (USDA, 1989a, pp. 17–18).

Third, the mode in which U.S. food aid is given has changed dramatically. In the early years of the PL 480 program, "sales" for local currency was the dominant mechanism. That mode gave way to long-term dollar sales in the early 1960s (new Title I). Perhaps less widely understood is the more recent quantitative change to Title II food aid that is distributed largely through nongovernmental organizations (NGOs). The 1989 foreign assistance request, for example, specified almost 60 percent of the total as Title II—a doubling of the proportion that was sent in this manner during the mid-1970s. This change

---

[1]Title I, under amended U.S. Public Law 480 (PL 480), refers to long-term dollar sales at concessional interest rates. Title II encompasses barter agreements and grant aid given mainly via nongovernmental organizations. Title III refers to dollar loans, which can be converted to grant aid if recipient countries agree to certain policy changes. To date, Title III has been quantitatively unimportant, and in recent years, U.S. food aid has been about equally divided between Titles I and II.

reflects in part the growing strength of NGOs in the lobbying and legislative process. It also means that more food aid is going in the form of emergency relief and in support of particular projects. From an aid-assessment point of view, this change has had a further consequence. There are very few academic studies of Title II, perhaps in part because the uses to which it has been put are difficult to assess and because few analysts have relished the thought of challenging the church groups and others that form the core of Title II proponents.[2] Whatever the reasons, there is little in the written record that provides an analytical basis for assessing Title II's success or failure with projects in recipient countries.

Finally, there is a growing maturity about the food aid literature, Title II notwithstanding.[3] The excessive rhetoric employed by both supporters and detractors of food aid has abated. Authors now seem to understand that food aid does not always create price or program disincentives, although it often does. Assessments of PL 480 have become less "theological," less "either/or," and more empirical in character. Many would argue that this maturity has been a long time in coming, but at least discussions about the future of food aid need not dwell on defending or demolishing overstated arguments of the past.

Questions about Food Aid in the Future

In assessing the future of food aid, some assumptions about scale should be made explicit. Since the mid-1970s, total food aid has ranged between 6 and 13 million tons of cereals annually. The variation, even year to year, has been substantial. Future discussions about level and stability should logically be linked to programmatic issues. For example, will emergency aid, in response to weather and climatic changes, need to be increased substantially? Can food aid provide the basis for greatly expanded food-for-work or human-capital programs?[4]

Although issues of program, substance, and absorptive capacity are interesting, the scale of food aid is much more likely to be determined by commercial interests of donors than by the need of recipients. In

[2]An important exception is the very critical study by Jackson (1982).

[3]One of the best recent volumes is Singer, Wood, and Jennings (1987). The August 1989 issue of *Food Policy*, which focuses on food aid in Africa, also contains a very useful set of essays.

[4]See earlier proposals, such as Thomas (1971) and Schuh (1981).

particular, total shipments of food aid have been directly related to the size of cereal stocks and inversely associated with the level of real cereal prices. Since the mid-1970s, for example, the R^2 between global food aid and global cereal stocks is .68, and the elasticity is .71. That is, a 10 percent change in stocks is associated with about a 7 percent change in food aid. Moreover, the R^2 between total food aid and U.S. food aid is .91, and the elasticity is .96.[5]

These relationships emphasize the effects of commercial policy generally, and U.S. policy specifically, on food aid. They also show one of the important self-limiting characteristics of food aid. During periods of low stocks and high prices, food aid is most important to the least-developed countries but least available from donor countries. Conversely, when stocks are large and cereal prices are depressed, donor countries make available larger quantities of food aid, thereby compounding problems of disincentives in poor countries. There are exceptions to the foregoing propositions, both among donors and recipients; nevertheless, they provide a sobering reality for all evaluators of food aid.

Uruguay Round

If food aid continues to be conditioned heavily by what happens in commercial markets, it also follows that food aid issues are intimately connected with the Uruguay Round of trade negotiations that are under way as this is written. This chapter is not the place for a full assessment of these negotiations; nevertheless, some discussion seems appropriate.

The priority question has to do with the overall success of the negotiations. Will developed countries reduce their agricultural output by phasing out price supports and other subsidies? If the answer to that question is yes, stocks of donor countries are likely to be lower on average, and real cereal prices might be expected to rise. At the margin, therefore, it seems that the more successful the Uruguay Round, the dimmer the outlook for increased amounts of food aid.

Completely unsubsidized, undistorted trade for agriculture, however, seems utopian at this time. If instead the result is a papering-over of certain trading problems between the United States and Europe, without altering the fundamental problems of highly subsi-

---

[5]Calculated from data in FAO (1989). Results from first-difference models are very similar.

dized domestic agriculture, stocks in developed countries might build, and food aid might increase.

Several countries, including the United States, have argued that "bona fide" food aid should be excepted from the trade-distorting issues that are at the heart of the negotiations.[6] Larger amounts of food aid could continue to be one of the escape mechanisms in the subsidy-reduction calculations, even in a largely successful Uruguay Round. The interpretation of "bona fide" might prove important, however. For example, would only food aid in the form of grants (as opposed to loans) be excepted from the calculations, and would it matter from which ministry the aid was offered? Such technical questions might also have a very large bearing on U.S. shipments, especially on the relative quantities shipped under various titles.

U.S. Agricultural Policy

Because the United States is such a dominant donor (56 percent of total food aid in 1988–89), what happens to its domestic agricultural policy is crucially important. Hazarding a guess on the effects of the 1990 bill is undoubtedly a foolish thing to do. Yet there are four broad forces at work that would seem to limit the medium-term buildup of cereal stocks. First, and to the extent that the executive branch has its way, there are likely to be downward pressures on loan rates for cereals. The secretary of agriculture, Clayton Yuetter, was a very strong advocate of keeping U.S. products priced competitively abroad when he was special trade representative.

To the extent that there begins to be a buildup of policy- or weather-induced surpluses, two other policy elements may become dominant, especially if the Uruguay Round is successful. Specifically, the use of marketing loans and the export-enhancement program are unlikely to go away in the 1990s. Whatever the rationale (or lack of one), farm groups are likely to push hard for these varying forms of export subsidies. To the extent that such subsidies prevent or eliminate cereal surpluses, they are likely to be competitive directly with food aid. GATT problems aside (and this is no minor issue), the discounted present value of cereals moved under such programs is still

[6]The negotiating proposal to GATT from the United States argues: "Certain types of policies would be permitted, and therefore, would be excluded from the aggregate measure [of subsidy reduction]. These policies would be either production and trade neutral or have such a small effect as to be inconsequential . . . [including] bona fide foreign and domestic aid programs" (U.S. Negotiating Group on Agriculture, 1987, p. 3).

higher than most forms of food aid. And though the United States provides food aid for many reasons other than simply to reduce stocks or maximize revenue, it would be naive to believe that the economics of alternate forms of international shipments are unimportant in U.S. policy decisions.

Third, the large and persistent budget deficit in the United States will eventually affect agriculture. The weather-induced rises in commodity prices in 1988 substantially lowered the cost of farm programs, even including drought payments.[7] Given other program needs and the limited number of options open to lawmakers, however, farmers may have an increasingly difficult time in securing support payments in excess of $15 billion or so, even if the gap between support and market prices increases substantially.

Finally, concerns in the United States for conservation and the environment are likely to become more influential in limiting stock buildups than in the past. The combined acreage of set-asides, paid acreage diversions, and the conservation reserve totaled about 50 million acres in 1989. In addition, concerns about mining groundwater to use for irrigation to produce "unneeded" surpluses are heard with increasing frequency, as are comments about soil erosion and various forms of agricultural pollution. An apparent requirement now for any paragraph on agricultural policy originating in Washington, D.C., is that it contain the word *sustainable*. Although the rhetoric of sustainable agriculture may itself not be sustainable, that the United States has currently chosen to idle more of its land than the combined *total* arable lands of West Germany, Denmark, Portugal, the United Kingdom, and of the Netherlands indicates yet another dimension of U.S. determination to limit future buildup of stocks.

Implementation of Food Aid Programs

The argument thus far has focused on the linkage between levels of donor stocks and levels of food aid and on a series of policy initiatives currently under way that might well limit in the future the size of stock buildups. If these propositions are correct, the general out-

[7]Expenditure on U.S. commodity programs were roughly $18 billion in 1985, $26 billion in 1986, $22 billion in 1987, $12 billion in 1988, and $14 billion (estimated) for 1989. There are currently about 2.3 million U.S. farmers, about one-third of whom produce about 80 percent of total output and receive about 85 percent of government payments. See USDA (1989b, p. 55).

look is probably for a global food aid program of approximately the same scale as now exists. Two final issues remain to be discussed. Is this result good or bad? Can food aid be used more effectively?

Food aid analysts probably should not bemoan the fact that global food aid is unlikely to exceed about 10 million tons in the foreseeable future. This is a personal judgment derived from two conclusions. First, there seems to be a maximum amount of annual food aid that is truly useful in the world for alleviating poverty and hunger. Second, quantities of food aid beyond that maximum tend either to have too many strings attached by donors or to be put into recipient countries when the policy environment mitigates against their successful use.

During recent years, about 15 percent of global food aid has been used as emergency relief, about 30 percent in support of projects, and about 55 percent as non-project aid (FAO, 1988, p. 37). The emergency relief component reached a high of 25 percent in 1985, during one of the worst years of the Sahelian drought, but in general, the proportion among the three uses has been fairly stable.

Emergency Relief

Recent historical experience suggest that 2 to 3 million tons annually for disaster relief is a reasonable estimate for the 1990s. This sum would be woefully inadequate were there a sustained drought in South Asia. It is unreasonably low also in the sense that many people will go hungry, even starve, in emergency or near-emergency situations each year. By and large, however, the donor countries have been generous in providing relief and have responded well when recipient governments have permitted external aid supplies. Unfortunately, all too many emergencies arise as the result of wars or internal political upheavals in which food becomes an instrument of oppression. In such instances, aid can be of little help, even if hunger is widespread.

The logistical efficiency of humanitarian relief seems surprisingly high. NGOs are at their very best in such situations and are capable of mobilizing grass-roots organizational support that governments typically cannot. Although some observers are distressed about the waste of relief supplies and failure of all the food to reach hungry people in remote areas, the real problem may be unrealistic expectations. Famines typically imply a complete breakdown of institutions within the food system. In such situations, if two-thirds of the grain reaches an appropriate destination, perhaps there should be delight rather than dismay. Whatever the problems with food aid, there should be agreement that its priority use is in emergency relief and that the global program is large enough to accommodate this form of assistance.

Project Food Aid

Food aid in support of projects is much more controversial.[8] Many of these projects are related to poverty alleviation, are oganized by NGOs, and, when judged narrowly, are quite successful. Such projects, at least for U.S. food aid, might be especially important in poor countries that are at political loggerheads with any given U.S. administration. Such projects often suffer from two fatal flaws, however. First, they are often so resource intensive—both human and financial—that they cannot be replicated. It may be possible to "save" a village, which in and of itself may be useful; however, in quantitative terms, these isolated projects can rarely be replicated into programs serving an entire country. Second, bad government policy toward food and agriculture typically destroys good projects. If policy discriminates against poor people, there may be little that individual projects, however well intended, can do to rectify widespread hunger.

Program Food Aid

Food aid in support of development programs is even more controversial than project aid. In principle, it should be possible to use incremental food resources to assist with structural adjustment and to improve the nutritional status of poor people. Moreover, it should be possible to design compensating fiscal policies that keep food aid from depressing farm prices or causing other program disincentives. There is even some evidence to suggest that donor countries are becoming more adept at solving some of the logistical problems that have plagued food aid in the past (such as the wrong commodity to the wrong country at the wrong time). Nevertheless, there are two features of program food aid that make its successful use self-limiting.

First, when food aid is offered in large quantities to support security interests, the record is clear that domestic food systems are seri-

[8]For a fuller description of these difficulties, see Jackson (1982) and Dearden and Ackroyd (1989, pp. 226–28).

ously damaged. There obviously can be honorable debate about whether short-run political or military objectives should dominate long-run food and agricultural interests, but the existence of such a trade-off needs to be acknowledged.

A second major difficulty involves the interaction between a recipient country's domestic food policy and its successful use of non-project food aid. Those countries with "successful" food policies seem to accommodate reasonable amounts of food aid fairly well. Indeed, such countries have typically gone on to become commercial importers of grain. The problem is that countries with inadequate food policies seem to need food aid more urgently. Too often, this inadequacy takes the form of inept macroeconomic policies, under-investment in agriculture, and excessively low agricultural prices to producers. In such instances, non-project food aid provides a "soft" alternative to policy reform, thus further hampering the food system. Conditionality requirements on policy reform might work in principle, but they have not worked in practice. A major constraint, in both the donor and recipient nations, seems to be a lack of competent food policy analysts.[9] The capacity to improve basic food policies or to accommodate large amounts of food aid successfully into policy regimes that are substantially less than optimal is not a task for bungling bureaucrats in either donor or recipient nations. Too often it has been.

Conclusion

There is no adequate answer in this comment or in the preceding chapter to the question about food that is the most often asked. If there are surpluses of food in some countries and deficits that lead to widespread hunger in others, why is there no simple way to work out food aid arrangements to solve both problems? One could lay out elegant theoretical schemes in an attempt to answer this question. But in the near future, the distressing conclusion must be that a series of political, economic, and technical constraints will prevent large-scale changes in food aid. A program of 10 million tons, bounded by many vested interests, thus seems the most likely forecast for the 1990s.

[9]A lack of analytical competence, for example, is likely to limit several of the innovative suggestions made by Mellor for using food aid to assist developing countries with policy adjustments. See Mellor (1988).

References

Dearden, P. J., and P. J. Ackroyd. 1989. "Reassessing the Role of Food Aid." *Food Policy* 14 (August): 218–31.

Food and Agriculture Organization of the United Nations (FAO). 1988. *Food Aid in Figures.* Rome.

——. 1989. *Food Outlook,* no. 6 (June). Rome.

Jackson, Tony, with Deborah Eade. 1982. *Against the Grain: The Dilemma of Project Food Aid.* Oxford: Oxfam Print Room.

Mellor, John W. 1988. "Food Policy, Food Aid, and Structural Adjustment Programmes." *Food Policy 13* (February): 10–17.

Schuh, G. Edward. 1981. "Food Aid and Human Capital Formation." In G. O. Nelson et al., *Food Aid and Development,* pp. 49–60. New York: Agricultural Development Council.

Schultz, Siegfried. 1987. "Food Aid: An Effective Instrument of Development Policy?" *Intereconomics* (May-June): 137–44.

Singer, Hans W., John Wood, and Tony Jennings. 1987. *Food Aid: The Challenge and the Opportunity.* Oxford: Clarendon Press.

Thomas, John Woodward. 1971. "Rural Public Works and East Pakistan's Development." In Walter P. Falcon and Gustav F. Papanek, eds., *Development Policy II: The Pakistan Experience,* pp. 186–236. Cambridge: Harvard University Press.

United States Department of Agriculture (USDA), Economic Research Service. 1989a. *World Food Needs and Availabilities, 1988–89: Spring.* Washington, D.C.

——. 1989b. *Agricultural Outlook.* AO-154 (July). Washington, D.C.

U.S. Negotiating Group on Agriculture. 1987. "United States Proposal for Negotiations on Agriculture." Washington, D.C.: Office of the Special Trade Representative, July. Typescript.

10

The Nature of the State and the Role of Government in Agricultural Development

Just Faaland and Jack Parkinson

The contributions a government can make toward modernization in agriculture is restricted by its powers, its stability, and its acceptability. The power of a government is reflected in its ability to ensure law and order, to enforce political and administrative decisions, to uphold the discipline of contract in economic life, and to provide for many other needs of society that require government action. The stability of government provides the basis for continuity and predictability of public action to which individuals, markets, and organizations can relate with assurance. It is not only a question of the longevity of particular governments but also one of the nature of the state. In the case of a well-balanced democracy, for example, continuity and predictability may be little affected by successive changes in government. It is the totality of the political system that provides continuity of the government function.

The acceptability of government conditions what it can do, what the response will be to its actions and pronouncements, and how widely and fully its policies will be adopted and adhered to. Acceptability is not to be taken as an attribute associated only with democratic governments; history is replete with instances of authoritarian governments, even fascist governments, that have enjoyed high levels of acceptability and therefore have had wide scope for influence in general and upon agriculture in particular.

Our goal is to seek the characteristics or attributes of the "nature of the state" that have a bearing on the role of government in agricultural development. Rather than attempting to give a systematic treatment of state organizations of different natures, such as feudal or authoritarian, religious or ideological, populist or democratic, we have concentrated on Myrdal's concept of soft or hard states, a characteristic that cuts across other differentiations on the nature of the state.[1] We have done so because we wish to bring into the debate a dimension, too often neglected, in the analysis of options as between alternative approaches for government action in economic development. Our experience indicates that decisions about the role of government are often taken as if the government were all-powerful and assured of full and automatic compliance with its decisions. This is far from the truth, of course, and decisions taken on such presumptions are bound to be ineffective or worse. Nevertheless, again and again, both in argument and actual policy making, this error is made. Moreover, irrelevant comparisons are made between options that are not in fact available—for example, between a perfectly functioning market economy and a perfectly executed command economy. When account is taken of the inevitable high degrees of imperfection in real-life situations, the comparisons of "ideal types" at best can have intellectual or pedagogical value. A rational decision must derive from a case-by-case examination of alternatives available and of imperfect systems and courses of action. A realization of the importance of this principle should make us highly skeptical of generalizations, for example, urging most African countries to open their economies more widely to the "free forces of the market," to remove subsidies or restrictions on agriculture or industry, or to effect other sweeping measures. The task of the policy analyst is much more demanding than the formulation of such generalizations: real and relevant assessment of options must be conducted in the specific circumstances that prevail in a given country at a given time.

In the following pages, we discuss different areas of action wherein a government might seek to intervene in the development of agriculture. The nature of that intervention often depends on the characteristics of the state itself—its tendency to retain power at the center and the degree of power it has to promote agricultural development.

---

[1] "These countries [the countries of South Asia] are all 'soft states,' both in that policies decided on are often not enforced, if they are enacted at all, and in that the authorities, even when framing policies, are reluctant to place obligations on people" (Myrdal, 1968, p. 66; see also chapter 18, sections 13 and 14 of the same work).

We reflect on the potential strategic role of agriculture in overall development; specifically, we address the question of the relative rates of expansion of industry and agriculture and the ways these sectors may interact. This discussion helps to set the scene for our subsequent analysis of the government interventions that might be appropriate—whether to be directly involved as an entrepreneur or to create the technical, human, and institutional infrastructure conducive to sustainable agricultural development.

The Basis and Scope for Government Action

Very little can be said about governments and the power of the state in developing countries to effect change that can claim some fairly universal applicability. Generalizations are fragile things, and, in the case of developing countries, there are likely to be many exceptions. Hesitancy must also be felt about attempting to generalize about the relationship between the nature of the state and agricultural development in such countries, for there are great variations in the forms and functioning of governments as well in the countries themselves. Yet can we not find sufficient similarities in the situations of developing countries for something useful to be said? Clapham's (1985) study of Third World politics provides a feasible basis for some generalizations. In his view, governments of Third World countries, and perhaps some developing socialist ones, have something in common in their existence: the result of being brought together in what is in many respects a single global society, economy, and political system. "Before the creation of a global political order, and the global economy to go with it, the infinitely varied political structures of Asia, Africa and the Americas possessed no common features which could bring them together as a single category for analysis. The importance of European colonialism is that it was, more than anything else, the means by which this global political and economic order was created" (p. 12).

This statement may be true in a global context, but colonialism has not conferred the full benefits of a system of government that is both accepted and responsive to the need to adapt to changing circumstances and political pressures—one that can assimilate new social forces and alignments which result from modernization. Powerful, stable, and accepted government is all too rare among developing countries, and even when attained for a time, it is liable to break down under pressure. Almost all of them have been the subject of insurgency, revolt, or military conflicts at one time or another, sometimes frustrating development or reversing it for many years, as we know from personal experience in the Sudan and Bangladesh.[2]

The Nature of Government: Power at the Center

The fragility of regimes struggling to maintain highly centralized power, without the support of many years of recognized legal authority conferred by the wide acceptance of goals and means, has major consequences. The first of these is likely to be reflected in the concentration of the efforts of ruling elites to keep themselves in power by sustaining their supporters and buying off the opposition, with little primary regard for what might be judged objectively to be the needs of the country. Strong leadership is eschewed in the interests of establishing support and consensus to a much greater extent than would occur in a more advanced society with strong democratic traditions and accepted procedures for the selection of governments. One result of the efforts to sustain support is that the character of government is strongly patrimonial, with authority ascribed more to personality than to office. This aspect extends downward from ministerial patronage to that of civil servants and to the lowest official who has the power to exact some favor or grant some exemption. Socially endemic, and comparatively acceptable, *backsheesh* or squeeze may give rise to bribery, extortion, and privilege as a means to secure access to wealth while bypassing the market. The economic repercussions of such entrenchment can be far-reaching. In Pakistan or Bangladesh, the system of physical controls adopted as a legacy of wartime planning, when the use of resources was determined by the state, was unthinkingly assumed to be of equal value in the fight for development. When this was demonstrably proved not to be the case, dismantling the apparatus of control was slow to take place: the system conferred upon officials a measure of power and the opportunity to exploit it. The effects were felt on agriculture no less than on industry.

In the Sudan and Egypt, the fear of political repercussions from unemployed graduates led to their being put on the government payroll even though they had little or nothing to contribute, thereby

---

[2]For an account of the incidence of conflict in developing countries up to 1965, see Huntington (1968, p. 4).

draining resources that might have been directed to agriculture. Such misuse of resources is perhaps minor compared with the extortions of other regimes, as in the cases of Uganda, the Central African Republic, or the Philippines, but the example points to the pervasiveness of the effects of efforts to extort tribute.

Still another aspect of the consensus state lies in the lack of accountability of officials to government and of government to the people of the country. To make officials answer for their actions, except when efforts are made to discredit them, is to disturb the elements of support and kinship that the regime needs for its coherence and maintenance of power. Officials are thus seldom brought to account and errors are unlikely to be corrected. Moreover, consensus itself dilutes the responsibility of individuals and groups.[3]

The fight to hold power at all levels is likely also to impede decision making by making people reluctant to step out of line. This tendency has colonial antecedents. Service under a strong colonial authority is not conducive to taking initiatives; decisions are likely to proceed according to the book and with suitable, and time-consuming, references to other authorities. There is little scope for initiative under such circumstances.

These proclivities are reinforced by the power-guarding nature of the regime. This tendency may be reflected in the stultification of activity on the part of ministers looking to the all-powerful president for leadership and not daring to move without it.[4] Delays inevitably follow, and economic activities dependent on speedy decisions are frustrated. But the effects are even more far-reaching. By definition, a strongly centralized regime is not interested in devolution and is resistant to movements in that direction. Power is jealously guarded. Local governments or authorities are strongly controlled and regulated; they are deprived of fungible revenues and forced to refer decisions to headquarters. Even a regime secure in its political power is likely to look askance at providing local bodies with revenues, including taxation powers.

Devolution is impeded in other ways. When power is centralized, civil servants are likely to be reluctant to move from the center. This aversion is, of course, reinforced by other disadvantages: the absence of good housing, educational facilities, and other trappings of civilization. This situation is of special importance for agriculture, the most

[3]For a more extended discussion, see Hydén (1988, pp. 20–22).
[4]For a discussion of the consequences of centralization of decision making in presidential hands, see Hydén (1988, pp. 20–21).

decentralized of industries. Competent civil servants able to supervise government policies and put them into effect are hard to find. In the absence of a local administrative presence, coordination between related ministerial activities is difficult and has to be resolved at the center rather than at the point at which the policies impinge, which would be most efficient.

Though power may lie very uncertainly with the government of the day, the civil service can provide a more constant form of guidance. In Bangladesh, governments and ministers come and go, but civil servants are much more firmly ensconced in position, and it is with them that much real power lies. An opportunity is given to civil servants to provide guidance, and they carry with them some of the more effective aspects of colonial rule, although they are unlikely either to wish or to be able to show much initiative for change. They too are affected by the uncertainties of government regimes and the power scramble, but they may have been able to maintain some traditions of service and to protect themselves from the struggle for power by the effects of seniority.[5] In practical terms of agricultural development, the lesson is clear: it is the civil servants that count most in many situations and upon whom policy determination as well as the implementation of projects often will depend.

The Power of the State to Promote Development

The picture painted above and the reasons why it has come about cannot, of course, fit the circumstances of all countries, but, with rare exceptions, governments of developing countries often seem to be ill-organized and to have only limited capacity for supporting agricultural development. The many complaints from aid donors of administrative failures, slowness of decision making, and inadequacies of economic policies often have their roots in the system of government that has grown up to replace that imposed, in many cases (although not, of course, in all) by an earlier alien regime, and by the failure to adapt it to the needs of rapid change. It is not within the mandate of outsiders, however, to alter the political system; they

---

[5]The attainment of a high-level position in the civil service is often determined at entry by performance in some examination taken many years before. Among those passing at a high level, this performance is generally recalled with great satisfaction, and there is an intimate knowledge by all concerned of their position in the pecking order. Of course, the disadvantages of promoting officials according to bygone performance and established precedence are obvious, but the system may have advantages too.

must understand this fact and accommodate to it. It is not generally the case that government procedures are unworkable; rather they have defects that have to be recognized, as far as possible, by devising suitable ways of proceeding while not expecting too much of the system.

The exasperation of foreign development officials with other aspects of malfunctioning government systems must also be contained. It is not very productive, for example, to complain that the entire administrative machine is riddled with corruption or that hardly a transaction takes place or a permission given without some cut being made. Very frequently, those in authority are so badly paid that it is only by finding some supplement to their incomes that they can continue to function. It is widely recognized that bribery, however reprehensible in Western eyes and negatively viewed as a suboptimal use of resources in the eyes of economists, is nevertheless a means of getting things done.[6]

In a wider context, it must not be assumed that policies that seem soundly based to Western economists are acceptable to those in power. The quest for economic efficiency must be weighed against other considerations as part of the desire to remain in power and carry out political objectives. The power of persuasion lies not so much in considerations of technical efficiency as in political gains and losses; if the failure to adopt policies conducive to technical efficiency would clearly be costly in loss of support and resources in the judgment of those ruling the country, the need for such policies may be accepted, but not if the reverse is true.[7]

Asian countries are not the only ones that can be put into Myrdal's classification of the soft state. Hard states are few and far between, and they will be judged more by the effectiveness of their policies than by their readiness to make unpopular decisions and see them carried through. Mozambique, once a hard state in its intentions to push ahead with its strongly socialist policies, has been nearly defeated by engineered political unrest. Tanzania, also bent on being a hard state, has found that hardness does not always pay. All policies should follow from the intention to attain objectives, but the objectives must be feasible in relation to policies that can be put into force and made acceptable to those affected by them.

[6]As Huntington (1968, p. 69) writes: "In terms of economic growth, the only thing worse than a society with a rigid, overcentralised, dishonest bureaucracy is one with a rigid overcentralised, honest bureaucracy."

[7]For a discussion of the consequences (often unpredictable) of policy changes see Thomas and Grindle (1990).

Although compared with the views of the 1960s, there may be much less faith in the powers of the state to promote development, there can be few who do not believe that the state has some role to play that transcends the establishment of justice, law, and order. Nevertheless, when planning was in its heyday, there was an implicit assumption that government had the capacity to formulate and execute detailed economic plans. Experience has shown otherwise. Third World governments could, and did, draw up plans, although many of them required the services of foreign experts to do so, but that was only the start of the operation. Execution of plans often presented considerable difficulties, partly because of limited administrative capacity, sometimes because the plans themselves were not well conceived, and in other cases because the basic assumptions of the plan proved to be untenable in a shifting set of economic circumstances.[8] Difficulties of implementation are particularly acute in relation to the private sector of the economy. It is difficult to reconcile the command economy of planners with the independence inherent in effective private enterprise and all too easy to equate planning for the private sector with repressive control. Thus, in many ways, planning is inimical to private activity, although it may be of help to it indirectly by providing infrastructure and establishing supportive institutions. Since a very large part of the economic activities of largely underdeveloped countries falls into the category of the private sector, and most evidently in a myriad of self-employed farmers, and in light of the poor performance of centrally planned economies, it is not surprising that planning is out of favor.

A decision not to rely on central planning does not mean that government has no role to play in agricultural development. It could, for example, attempt to intervene directly by establishing state farms, although, as we explain below, we think that this move would often be unwise. What is really needed is limited and carefully selected measures to assist farmers while relying largely on individual self-interest to accomplish an agricultural transformation. The function of the state mainly consists in creating conditions under which agriculture can thrive. Above all, it needs to refrain from actions that are needlessly damaging to agriculture, including serious distortions of the incentive system.

[8]The first five-year plan of Bangladesh provides an example of a plan that was out of date even before it started to be executed; the rise in oil prices and the price of other imported materials was so great as to invalidate the estimates used in the plan of the amount of imports that could be afforded.

The design of supporting policies for agriculture must often proceed by trial and error—in other words, learning from mistakes—so that administrators may fight shy of it. If, however, it is accepted that government interventions designed to increase agricultural output are necessary, careful consideration must be given to the instruments that can be used and the measures that have some chance of acceptance and success. The government invariably has only limited resources for agricultural support, and policy makers must be selective in what they advocate. Key areas for intervention need to be determined, and it is to be expected that they will vary considerably from country to country, although increased use of modern technology is almost certain to be a crucial ingredient. In some cases government action in one key area may be sufficient to promote higher productivity using new technologies. The provision of irrigation facilities, by promoting double cropping, may be all that is required or, if complementary inputs are necessary, sufficient to actuate a suitable response on the part of farmers in the sequence discussed by Hirschman (1958). In other cases, greater government involvement may be needed when the adoption of new technology requires several ingredients to reap full effects of synergy.

A Role for Government in the Agricultural Sector

In the discussion that follows we address ourselves largely to consideration of the least-developed countries. For them the essence of the problem, as we see it, and therefore the context in which the nature of government intervention needs to be considered, is that agricultural output in most developing countries is too low and its rate of increase too slow to support more rapid development. As a result, employment opportunities cannot be expanded on the scale needed, investment is held back, and consumption suffers. Underlying these considerations is the need in many developing countries to reduce hunger by ensuring that the poor have the means for an adequate diet.[9] Experience since the mid-1960s suggests that it is more difficult in developing countries to increase agricultural output than industrial production. A glance at the record for the poorest countries for the period 1965–80 shows that the average increase of agricultural output for the group was less than half of that achieved by

[9]For a discussion of these issues, see Timmer, Falcon, and Pearson (1983).

industry.[10] For the period 1980–86, the rate of increase was much closer, but only because China recorded a large increase in agricultural output, thus swinging the average agricultural output for the group closer to that of industry. It is noticeable that in the case of individual countries, when the rates of increase in agricultural and industrial output are nearly the same, or the relationship is reversed, this result is not because of success in agriculture so much as failure in industry. In the long run, it would not be expected that the growth of agriculture and industry would keep pace because of the operation of Engel's Law, and it is also true, of course, that industry, starting from a low base in most instances, might be expected to show much greater increases in output in the initial stages of development. Nevertheless, the demand for agricultural products in the early stages of development, if it can be satisfied, increases at much the same rate as the increase in income that is achieved.

The impediments to expanding agricultural output are many and varied. Climatic and soil conditions are often unfavorable, but even when this is not the case, many other obstacles must be overcome. In industry, production is concentrated in comparatively compact areas, which can be made fairly easily accessible. In primitive agriculture, most forms of cultivation are extensive, and many of the areas of cultivation may be remote—for example, in the hill tracts of Laos or in the west of the Sudan. Even areas that are not widely dispersed may be hard to reach because of lack of transport facilities.

Another issue is that of control. In industry, control of production operations is usually centralized within the plant; control is in few and, by ordinary standards, expert hands. The workers do not need to be familiar with production processes or systems of industrial organization to contribute to production; they are not managers except in the simplest sense. It is in the nature of large firms that they have access to technical knowledge and finance. They frequently are backed by the resources of multinationals and can call on their central organizations for assistance. A developing country, always assuming it is prepared to call on foreign assistance, can obtain turnkey plants, probably less efficient in their operations, than those in the developed world and sometimes with negative value added, but nevertheless functioning and producing.[11]

[10]For details of the figures see World Bank (1988, p. 224).
[11]See Balassa and associates (1971, p. 239) for examples of negative value added, including, for Pakistan, sugar refining and the edible oil industry as well as motor vehicle assembly.

In agriculture, ready-made innovations cannot be forthcoming in this way. It is not, for the most part, acceptable to invite foreign firms to establish large farms even if these could function efficiently; the existing farming arrangements cannot be easily disturbed without social disruption. It is hard to transfer modern technology and all the inputs that are needed to go with it quickly. If it were not for these and other impediments, it would be possible to increase agricultural output greatly, although not to the extent and with the assurance that modern corporations could transform the industrial scene. A note of caution should be added. Early postwar efforts to transform agricultural output in some of the British colonies in Africa, which might collectively be referred to as the groundnuts schemes, showed the shortcoming of large-scale agricultural operations starting up in ill-reconnoitered environments.

Government as Entrepreneur

The most direct way in which the government can act as an entrepreneur is to establish state farms, but a decision to do so may have very little relationship to considerations of economic efficiency. The establishment of state farms in the USSR followed from a series of political decisions, designed above all for the achievement of political aims. "It goes without saying that if capitalism could develop agriculture which is everywhere lagging terribly behind industry it could raise the standard of living of the masses. . . . But if capitalism did these things it would not be capitalism: for both uneven development and a semi-starvation level of existence of the masses are fundamental and inevitable conditions and constitute the premises of this mode of production" (V. I. Lenin, *Imperialism the Highest Stage of Capitalism,* p. 59, quoted by Griffin, 1979, p. 178).

The subsequent social and economic consequences for the Russian people have turned most agricultural economists in the Western world against state farms as an instrument for agricultural production. The case against them is generally based on the unsuitability of officials for direct business operations. Government officials are not practical men, they are not profit-conscious, their training, attitudes, and preoccupations are not conducive to managing enterprises and so forth. Agricultural operations also do not lend themselves to systems of remote control. Effective farm management and labor supervision ensures the timeliness of operations, personal tending of plants and animals, maintenance of soil fertility, responsiveness to changing prices and techniques.

Lack of understanding of the requirements of agricultural operations and how to organize them was one cause of the failure of the attempt to achieve a sudden transition to state farming in Russia. There seemed to be an implicit assumption that to command was to accomplish. The idea that the influx of 25,000 urban activists to act as supervisors, farm chairmen, or political officers could be conducive to improvement in farming practices and performance appears ludicrous in retrospect. "Their ignorance of rural questions and misunderstanding of the peasant mind contributed to the excesses of the period" (Nove, 1988, p. 181). It was the failure to understand the peasant, or any other producer's mind, that really mattered. This shortcoming had been evident far earlier in efforts to secure supplies by state procurement at unremunerative prices.[12] Better to withhold grain than to sell it at prices that made sales unrewarding; better to kill the animals and salt them down than to allow them to fall into the hands of state farms. Eventually, circumstances forced greater understanding; rectification of the famine conditions of 1933 depended on giving the peasants more control over their own production. Of course, the outcome might have been different if the creation of state farms had been done more slowly and with proper regard for creating conditions that would be conducive to running farms efficiently. But the lesson remains: it is not to be expected that the circumstances of the Middle Ages can be transformed by decree or overnight, as Stalin seemed to assume; at most change can be effected only by degree.

In many respects, China seems to have escaped the potentially destructive effects of attempting collective farming. In the case of China, the transition to state farms should not be regarded as destructive, but the evidence seems to point to the view that success in recent years in moving away from the concept of comprehensive collectivization has brought even more productivity. Before 1955 there were 100 million family farms in China. Transformation of these via cooperatives to some 55,000 communes was not without its successes; in any case, controls via the command economy were relatively weak.[13] Nevertheless, recent progress seemed to hinge on two decisive acts of policy: the decision to increase agricultural prices by

---

[12]For discussions of the role and importance of agricultural prices as they affect production and consumption, see Timmer, Falcon, and Pearson (1983).

[13]See World Bank (1986, p. 104) for an account of Chinese agricultural progress.

substantial amounts in 1979 and the effective handing back of land to peasants to farm. By 1983, 95 percent of farm households were managing their own plots under contracts from collectives. Many households were given the right to manage these plots for fifteen years, thus encouraging them to make lasting improvements. As a result, grain output increased from 305 million tons in 1978 to 407 million tons in 1984. In a considerable measure, this improvement in output appears to have taken the form of a once-for-all efficiency gain although the trend of total output still appears to be moving upward. Nevertheless, the remarkable progress that was made contrasts vividly with the sluggish agricultural economic performance of other low-income countries over the same period.

In principle, there is no reason why state farms should be large, but it may be suspected that there will be a tendency for them to be so, if only for economy of administration. In many instances, however, size carries no advantage because economies of scale are not significant. Moreover, small farms in developing countries may often (though not invariably) make better use of land; the larger farms in developed countries operating many hectares with few workers would gain little in economy of machine hours if they were bigger, and even the smaller farms in developing countries can be accommodated with tractors of much smaller size than those in the tractor parks of state farms of many thousands of hectares.

In spite of convincing arguments against state farms as a first choice for agricultural development, there may be cases when they might be justified on grounds of economic efficiency as well as political or social acceptability. Plantations have shown their economic worth; large farms operated by settlers in Africa and elsewhere have demonstrated how to mobilize modern technology and management skills.[14] The arguments that such innovations could not be done by state-owned corporations are not entirely convincing. Moreover, sometimes a partnership between state and private corporations works well, as in the case of the mining corporation Debswana, owned jointly by Botswana and De Beers. The managing agency may be an outdated and highly criticized institution, but its efficiency might be greater in many respects than that of peasant agriculture,

[14]Binswanger and Rosenzweig (1986) argue that successful plantation operations are those for which there are either economies of scale in processing or marketing or considerable capital requirements in the case of crops with long gestation periods or, of course, a combination of these factors. This may, however, be changing as a move to contract farming gathers momentum to secure supplies, as Kirk (1987) has described with particular reference to transnationals.

particularly since the managing agency is more likely to be in a position to finance the means of production needed to exploit modern technology.

One of the most interesting alternatives to the state farm is the organization in the Sudan built up in the Gezira in a comprehensive framework, which in effect serves several economic, social, and political objectives and is operated by the Sudan Gezira Board with over 2 million acres under its control. As is well known, the scheme was launched as a result of British initiatives to provide, among other things, a further source of cotton for the Lancashire market in a profitable manner. The enterprise took the form of a joint undertaking between the British government in the Sudan and the Sudan Plantations Syndicate, and it was an early example of joint enterprise. In contrast to the Soviet experience, the scheme was mulled over for a long period; it was no hasty or ill-considered conception. Moreover, it was devised with an eye to political realities. Land required for the operation was acquired on long leases by the government. The profits of the operation were to be shared between the operating syndicate and the government. The land was let to tenants. The decision not to operate with employed labor seems to have been related to earlier difficulties in attracting workers willing to be employed.

This is not the place to recount the complicated and checkered evolution of the scheme following the signing of the agreement with the syndicate in 1919, but some points of particular relevance to the role of the state in agriculture are worth making. The Gezira scheme could not have come into being without state intervention. Without the authority of the government, land for it could not have been mobilized, nor is it likely that the Sennar Dam, on which the success of the project hinged, could have been built without a similar government initiative and recourse to the treasury in London.[15]

The structure of the Gezira operation, based on tenant participation subject to central control, has been shown to have disadvantages in spite of changes in the details of the operation; deterioration in much of the structure of the irrigation system over time threatens its operations in the future. For a period, the Gezira operation worked well and illustrated that there was some potential for socially profitable government intervention in agriculture on a large scale.

The Gezira initiative would have been impossible without large tracts of land to cultivate which could be operated as an integrated

---

[15]For an account of the early history of the Gezira operation, see Gaitskell (1959).

whole. It is clearly an inappropriate model for state intervention in the land-scarce territories of the Indian subcontinent or China, nor is it easy to see how other aspects of the scheme, such as the control of the production patterns of relatively large tenants, could be effected. Some part of the Gezira's comparative success must also be attributed to having one main initial objective: the production of cotton, a single crop. Other crops are now grown, and this diversification has added to the management problem. A control structure extending over more than 2 million acres of varied crops could well prove to have uncertain flexibility as farmers need to change crop rotations promptly in response to changes in market demand. One of the least satisfactory aspects of the Gezira scheme in recent years has been its marketing activities, which may also reflect its size and structure. If it were to be replicable in other conditions, where large tracts of land were readily available, experienced management would be essential. In the case of the Sudan, the colonial connection was used to provide qualified production managers, although the management of marketing proved to be a weakness.

In the unusual circumstances of post-independence Mozambique, state farms were established both in response to ideological promptings and as a means to keep the large abandoned Portuguese farms in operation. The role of state farms was planned to be more extensive, however; they were to provide support to neighboring small-scale agricultural operations and to cooperatives. This approach highlights a general problem: if the state regards it as its duty to try to improve agricultural performance and support the operations of small farmers, how best can this be done? The Gezira pattern as we have seen, provides one answer, but it is different from the contribution that might be made by state farms operated on the basis of paid labor. Where mechanization is important, access to a tractor pool might offer some of the economies of scale that would make it possible for impoverished small farmers to increase their output. Similar considerations might also apply to other items of equipment and to inputs such as fertilizers and pesticides. A nucleus farm operating in this way could also be a source of knowledge about modern cultivation methods. How far it would be possible for state farms to combine such a duality of function without a conflict of interest is uncertain. In a socialist country, such a function could be planned and allowed for in the state farm's accounts; in a strongly commercially oriented setting, this role would be much more difficult. Since, as we have suggested, state farms are likely to be big and therefore few in number, the catchment area might be somewhat limited. Greater extension of influence might be possible if, as was intended in Mozambique, agricultural cooperatives could also be drawn into the sphere of influence of the state farm, although here again the extent of such influence is bound to be limited by time and distance.

Another aspect of this extension of influence (which could be the province of state farms, parastatal farming operations or, and perhaps better, commercial farming operations) might lie in nucleus farming operations bent on securing supplies from a number of small, nonintegrated producers. The essence of such arrangements is a marketing facility that can allow the purchaser of the produce to secure output of good quality, produced competitively in assured amounts. This approach opens up a host of opportunities, ranging from the operations of independent and perhaps foreign firms to government marketing boards.[16]

Although state farms might exercise favorable influences on small-scale producers in some circumstances, perhaps the stronger possibility is that their activities may be harmful to farmers in their vicinity. Giant mechanized farms created in Egypt with Russian capital used large amounts of scarce land but provided practically no employment for the abundant labor. In Pakistan Punjab, large-scale agriculture led to eviction of tenant farmers and to serious social and political problems. In Iran, under pressure from the Shah, a gigiantic project in the Khuzestan Valley, including the construction of dams for power and irrigation followed by heavily subsidized tractors, gangplows, and combines, led to one of the largest "enclosure movements" in the twentieth century, accompanied by major political unrest.[17]

Intervention through Infrastructure and Services

State farms lie at one end of the possible gamut of government activities related to farming and represent considerable involvement. At the other end lies some minimum of state activities that needs to be undertaken but is unlikely to be carried out by the private sector

---

[16]It is significant that the policy of pouring scarce resources into Mozambiques's state farms has been abandoned, while small family plots are to be the backbone of rural development, following decisions taken at the Fifth Congress of Mozambique's FRELIMO party. See Maier (1989).

[17]We are grateful to Frank Child for a private communication outlining criticisms of large-scale state farming.

of the economy. Where should the balance lie between what must almost certainly be done by the state and what might be advantageous for it to carry out?

Physical Infrastructure. The clearest example of the need for state involvement in helping the agricultural sector lies in the provision of physical infrastructure, as we have already seen in the case of the Gezira. Even here, there are degrees of involvement, and measures may be directed to a wide range of objectives, including agricultural interests. Infrastructure as it is currently understood need not be provided by the state. The railways opened up the virgin lands of America, but, as in many other countries, they were privately constructed. In developing countries, the provision of expensive communication systems is almost certain to be beyond the capacity of the private sector to raise the money needed for construction, and lack of legal rights is likely to be a further impediment. If roads or railways must be constructed in order to open up agricultural land or enable produce to reach markets, it is almost certain that the state has to get involved. This statement does not mean that every meter of road has to be publicly constructed. If the key roads are in place, there may be sufficient incentive to construct feeder roads, or, as in the Sudan, there may be opportunity to move produce to market by using tractors for the secondary task of haulage. In other cases, if sufficient feeder roads have been provided from farm to pickup point to reduce the distance to a mile or two, the journey to market might begin in the form of a traditional headload. Somewhat similarly, it may be sufficient for the state to promote small-scale irrigation by ensuring a supply of credit and the availability of imports. But if such measures are not enough—as in locations where large dams must be constructed to ensure the supply of irrigation water—it is almost certain that the state has to be heavily involved. Projects such as the Indus water-replacement schemes and the construction of dams the size of Mangla or Tarbela suggest that it is not sufficient for the government or governments locally involved to attempt the work; international action involving other governments, possibly operating through international agencies, would be called upon. Even when there is a prospect of raising funds locally, it may still involve the use of government guarantees, thus requiring it to recover its costs as best it can.

Education. Many of the measures that the state needs to take to further its agricultural aims are likely to have beneficial effects far beyond the agricultural sector yet be of vital importance to encouraging agriculture and improving its prospects. Education is such a case, starting with the provision of universal primary education. Education is needed to provide such benefits as the accumulation of knowledge, the application of knowledge, the ability to calculate, the opportunity to place sellers and buyers on an equal footing, the provision of political awareness, and the ability to participate in institutions conducive to improved farming practices (such as cooperatives) and use of extension services and to seek protection from the law. Schultz was probably right in his basic contention that the peasant agriculturalist had perfected his use of known technology over the years, but the rate of change of technology is now so rapid that adaptation to it must be sped up and applied from first principles rather than imbibed over tens of years or centuries of individual experimentation and assimilation (Schultz, 1964).

In a similar way to education, agricultural performance is helped by other measures to improve human capabilities—by improvements in health, reduction in family size, and more efficient self-provision of services and improvement in living conditions. All of this requires the accumulation of knowledge and its application.

Research. In a simple sense, the increase in output per capita over the centuries has been made possible by the accumulation of technical knowledge and its application by various means. In agriculture, as elsewhere, technical progress, at first gradual, has expanded rapidly in the present century. It has not been the result so much of accidental discovery as a purposeful search for new ways of producing things or the substitution of new products to meet existing needs. These are not tasks that can be performed by peasant agriculturalists. In developing countries, they are also unlikely to be undertaken by indigenous business interests anxious to devise new technology as a means to sell new products, as they are in developed countries. Most developing countries do not have large corporations or multinationals equipped with research laboratories, large numbers of trained scientists, and the financial resources to underwrite research that only occasionally will come up with a worthwhile and salable product that can be promoted through well-developed sales outlets capable of imparting the technical expertise needed. If such research is to be carried out for the benefit of developing countries, alternative means of promoting it must be achieved.

There is no invariable pattern for establishing this research capacity, but in many cases, if not all, it is likely to require state intervention, in one form or another, to set up or support research institutions or agricultural universities. It is not necessary, of course for research always to be undertaken specifically for the needs of developing countries. Very often some new development, made first in a developed country, proves to be applicable in one or more developing countries depending on their particular resources and needs. Hybrid maize is a case in point. Developed in the United States in the 1930s, it proved to be suitable, in various forms, for other countries, although it has not achieved the success in Africa or Asia that was hoped for. The development of high-yielding varieties of wheat also owes much to American initiatives that involved the Rockefeller Foundation and the Mexican government and led to the establishment of CIMMŸT (the International Maize and Wheat Improvement Center) in 1966 (Pearse, 1980, p. 39). The International Rice Research Institute was set up in 1960, financed by the Ford Foundation, the Rockefeller Foundation, and several Asian governments, with the active participation of the government of the Philippines (Pearse, 1980, p. 40).

Since those days, other international institutions have been created, but there has also been increasing support for national research. Farmers tend to adopt new varieties first in the places where they are discovered, and national research centers, as opposed to international institutions, thus might spread the advantages of new seeds more widely and rapidly. Clearly this is an advantage, but in many cases the work of national and international research organizations needs to be differentiated. National research may best be concentrated on problem solving and application, because few developing countries can mobilize a critical mass of scientific talent, finance research adequately, and provide conditions suitable for scientific work. International centers have an advantage in these areas.[18] Thus many international institutions have been created to conduct research into a large variety of products, often with a regional association or with particular growing conditions in mind. They fall under the scrutiny of the Consultive Group on International Agricultural Research (CGIAR), which provides some measure of coordination of activities.

[18]The complementary nature of these activities should be noted. The greatest beneficiaries of international research are countries with strong national programs for research and its dissemination. See Horton (1986).

In addition to research into crops and livestock, improvements of farming systems are the subject of study.

There still remains, of course, considerable research to be done to obtain further improvements in agricultural output. Where the revolutionary discoveries have already been made, there may be less scope for further progress. It may be remembered that there are biological limits to increases in yields. But there still remain many products of importance to developing countries for which research could enhance prospects for increases in yields and improvement in quality, and there are other aspects of research, such as nitrogen fixation, that need to be encouraged. In research, a clear role can be recognized for the state, both at the national and international levels, and for other bodies with resources and vision, such as foundations. In countries at an early stage of development, agricultural producers are almost never in a position to organize such things for themselves.

Extension. It is universally recognized that research is of little use in agriculture without accompanying arrangements to disseminate its results. In a developed country, the spread of new technology is not the problem that it is in a developing one. If discoveries have been made by commercial organizations, they can be relied upon to mobilize their efforts to reach potential consumers. The media may well take up developments of particular interest and wide application. Trade journals also provide an instrument for dissemination. All such means are matched with a fairly wide receptiveness to new ideas and a readiness to experiment to some degree and to secure the necessary resources.

Even with all these obvious advantages, developed countries have felt it desirable to supplement the pressures and opportunities to innovate. With the intention of promoting change in agriculture, government officials have set up bodies such as extension services to help disseminate information, introduced subsidies to promote various actions, and provided educational facilities—and these are only some of the measures that have been attempted. For developing countries, the task of promoting change is much more daunting. To begin with, there are many more producers to reach in relation to the size of the country or its population and many fewer resources with which to accomplish it. In Bangladesh, for instance, it might be desirable to reach perhaps 15 million agricultural producers and provide them with practical information. By contrast, in England, with a similar area and half the population of Bangladesh but a much greater agricultural output, there would be about 300,000 producers, many of them with large outputs and most of them capable of obtaining the technical information they need without government assistance.

Any approach by the state to providing extension services must recognize that it is impossible to reach all, or indeed many, farmers. This constraint might be overcome by concentrating attention on model farmers who can act as exemplars for those in their vicinity. State farms, as we have seen, might be able to play a similar role, with the advantage that their workers can actively demonstrate the use of new technologies. There cannot, of course, be any real sanction on the activities even of model farmers, who may or may not be prepared to follow the advice of the extension worker or to grow the crops that the agent might wish to demonstrate. A possible solution to this dilemma might be for the state to involve itself more in cultivation, but on a small scale, designing its program to spread demonstration plots as widely as possible in various localities.

In spite of the best efforts of extension services to reach farmers, they are likely to have only a limited effect. The shortcomings need not prevent innovations from being adopted eventually by a large number of producers, although the process would be lengthened. Innovation also spreads in an informal way: a visit of one farmer to another results in the transfer of some promising seed for trial, or the information is exchanged wherever farmers meet to talk. The lessons of technology may be more easily appreciated if they come from another farmer actively engaged in rural life and fully aware of the various constraints that inhibit the adoption of new ideas.[19]

Cooperatives. Historically, the cooperative movement was seen as a way to protect the poor from the exploitation of capitalists, perhaps at the expense of some of the benefits that the capitalist structure could confer. The movement can perform this role in agriculture today, for example, by arranging the purchase of various inputs. A cooperative may also be regarded as a form of business organization with features that distinguish it from more conventional forms, such as partnerships or public companies, designed, perhaps, to achieve economies of scale in purchasing, processing, or sales that would not be feasible without cooperative action.[20] Cooperatives can also

---

[19]For an account of the spread of new technology in the Punjab and its greater repercussions, see M. S. Swaminathan's statement in Faaland (1982, pp. 92–93).

[20]See Parkinson (1965, p. 13), a report on the place of cooperation in the future development of agriculture in Northern Ireland.

improve farming practices by encouraging farmers to adopt new methods and to gain strength by acting in concert. This aspect of cooperative activity is of concern at the present time. The establishment of cooperatives, for whatever purposes, requires some form of legal framework, which is one of the functions of the state that only it can undertake. But the state's involvement may go much further, extending to the establishment of banking facilities and institutions designed to cater to the specific needs of the movement and its constituent societies. The state may be further involved by regulating the movement, to make sure that it operates within the law, that correct accounts are kept, and that the societies are solvent. In practice, the societies are unlikely to be self-establishing on a sufficient scale; promotional activities may have to be undertaken to get the movement operating effectively over a wide area.

The various viewpoints that students of development have taken on the functions of cooperatives and their power to exert a favorable influence on agricultural development tend to depend on which countries they have studied. The cooperatives in Denmark were clearly powerful and effective instruments for agricultural efficiency and progress. In other countries the picture is more mixed. In Bangladesh, cooperatives have not turned out to be the driving force for progress that it was hoped they would become. The Comilla model, established by Akhter Hameed Khan in East Pakistan in the early 1960s, was expected to provide the means to transform the agricultural sector and rural society by operating in a much wider context than simply agricultural development. This expectation stemmed from the view that progress depended on bringing supporting elements together in a mutually reinforcing manner. The Comilla model was a forerunner of the concept of integrated rural development as it was subsequently formalized in Bangladesh and promoted in many countries. It is probably safe to assume that the activities of integrated rural development projects have stimulated rural development in many aspects and helped hasten agricultural development, but the effects of such operations are hard to measure. Neither the cooperatives nor the associated wider activities have had, and perhaps could not have been expected to have had, the results that were hoped from them. Cooperative activities seem to have been the most disappointing aspect of this composite package.[21]

[21]For some critical reviews of integrated rural development programs and their results, see Ruttan (1984), Gower and Vansent (1983), Khan (1979). The experience of the World Bank is reviewed by Donaldson in Chapter 6 of this volume.

In other countries, the effectiveness of state-promoted cooperatives is also somewhat in doubt. The experience of Tanzania, with a long history of cooperative activity and development in the 1970s centered on villagization, again raises questions about the benefits that can be realized from systems of development based on communal interests. In Mozambique there has been no opportunity to test the power of cooperative activity to generate sustained agricultural development, although in the early stages of the cooperative movement such power seemed to have been established in an effective way in a limited number of centers and for a variety of purposes.

Land Use. Availability of land, and the right to use it, varies greatly from country to country. The state is responsible for setting the legal framework regulating the use of land, and it must provide control with a view to reconciling conflicting interests and promoting efficient use of land. At one extreme, there are countries with great scarcity of land in relation to their populations. Among the problems arising from this low ratio of land to population is the threat to the availability of agricultural land from the need to accommodate an expanding population, to provide for roads and airfields, and to develop towns and industry (invariably on the best agricultural land). In land-scarce countries, the government needs to give some thought to physical planning if it is to preserve land for agricultural use. Inequality in landownership is also a matter for concern under such circumstances, because it seems that as a general rule small plots of land are better managed than larger ones. But there may be reason for concern if there is a tendency for plots of land to become even smaller, more scattered, and less easy to work as continuing increases in population fragment ownership and reduce holdings well below a subsistence level. At this point the process of fragmentation may be reversed, with distress sales serving to augment the holdings of larger landowners. How the state can handle such a situation is not self-evident. Land reform can be a brittle weapon. The size of holdings can easily be disguised, and the process of reallocation, even if it can be enforced, is cumbersome and liable to disrupt production as it is carried through. When population pressure on land is extreme, land reform may be a continuing need as the forces leading to the loss of land by the very poor and the accumulation of it by the very rich are not interrupted by redistribution, only put back in time. Also, questions of compensation are difficult to resolve. Should compensation for loss of land by those well endowed with it be paid, or should the land be, in effect expropriated? If compensation is paid, would this payment contribute to future accumulation of land by those who have surrendered it compulsorily? Altogether, any government attempting land reform has a formidable problem on its hands. Even so, in some cases (for example, Korea and Japan), the results achieved amply justify the political risks and administrative difficulties that have to be faced. In most cases, however, efforts at land reform have not achieved a great deal.[22]

At the other extreme, relatively plentiful supplies of land, if unmatched with well-defined rights of usage, can be damaged or destroyed because the absence of property rights gives no incentive to manage the land with conservation in mind. It may be mined and abandoned as cultivation is moved to virgin territory and the process repeated. There are fears that this loss of soil fertility may be taking place in the Sudan with the expansion of extensive farming. When there is evidence of such land degradation, there may be no alternative to legislation designed to protect future fertility of the soil.

The Environment. Protection of the fertility of land is only one aspect of the more general issue of conservation. Only comparatively recently have the full dangers of ecological disaster been recognized as requiring not only national action but international intervention. The destruction of the rain forests, gaps in the ozone layer, and the greenhouse effect cannot be solved by the isolated actions of individual nations, and it cannot be assumed that a solution to them will be found by international accords to reduce the level of pollution being generated without other supporting measures. Developed countries are reluctant to raise the resources needed to combat pollution, although the seriousness of the situation is gradually making them more amenable, but developing countries are bound to be not only unwilling to find resources for the purpose but, in many cases, unable to do so. The need for international action to support antipollution measures in developing countries is becoming apparent. The ramifications of these international concerns are likely to be considerable. To take but one example, the energy needed to sustain development and an increasing population in South and Southeast Asia cannot come from indigenous fossil fuel resources; a solution might be to import fossil fuels from other countries such as the Soviet

---

[22]See de Janvry (1981) for an account of land reforms in twenty countries; see also Bromley (1981).

Union, the United States, or China, where they are in more plentiful supply. But because these three countries are likely to be among those most seriously affected by climatic change resulting from increased carbon dioxide, they may not be willing to develop an export trade in fuels, preferring to restrict production to limited quantities of coal for their own use (Revelle, 1982, p. 72). The implications for the developing countries in the region and for their agricultural operations—which, like anything else, require the use of energy—are clearly considerable. Here again, international action may be called for.

Other aspects of pollution generated in the course of agricultural development are not likely to be so global as those indicated above but may still have international repercussions. Preservation of wildlife, particularly in Africa, is an example. In Botswana, to take a relatively innocuous example, the increased number of cattle impinges on the habitats of wildlife and may lead to a diminution in their numbers or even, in some cases, to the threat of extinction. Although the measures needed to deal with this phenomenon are essentially internal to the countries concerned, they are also of concern to those countries anxious for the preservation of wildlife. One way such countries can express their concern is to make provisions in aid programs for preservation of the environment.

Aid may itself be a cause, even if unintentional, of environmental degradation. It is increasingly being recognized that many aid programs concerned with agricultural development were actually or potentially damaging to the environment. The remedies are clear in principle, but it may not always be easy to predict the consequences for the environment of various projects, and there may still be some nasty surprises in store, as evidence accumulates of harmful, but earlier unsuspected, effects of various activities.

A Facilitative Role for Government

The arguments advanced here point to a wide-ranging, yet circumscribed, role for the state in agricultural development in the Third World. The limitations placed on the role that can be played by government are essentially threefold. First, we believe that many governments in Third World countries have the characteristics of what Myrdal terms "soft states": weakness, insecurity, and reluctance to enforce policies. Second, of course, even strong governments find it hard to influence development constructively and effectively in such a widely strung operation as agriculture. Third, we have an inherent faith in the ability of people to improve their condition if given reasonable support in the form of services that the state, and often only the state, can provide. The precise role of government will depend, as always, on particular circumstances; yet some generalizations may be suggested.

It is undesirable for the state to involve itself directly in agricultural production unless it is very sure that it has the capacity to make its entrepreneurial activities a success. This is not generally the case. Instead, the state should attempt to provide the infrastructure that is conducive to agricultural development, but such infrastructure cannot be too widely spread to begin with if it is to be sufficiently concentrated to be effective. In this context, infrastructure has to be widely defined to include not just the provision of physical structure, such as roads, but also actions designed to make people more effective as producers and citizens, by way of education, among other things. We judge that expenditure on infrastructure for support of agriculture is often more cost-effective than expenditure within the agricultural sector proper.

The nature of governments in most developing countries is such that only limited direct influence can be exerted on agriculture. This points to the need to improve market mechanisms as much as possible. Prices in most LDCs tend to discriminate heavily against agriculture, and there are strong arguments for trying to reduce this bias. Nevertheless, in developing countries market forces alone cannot be relied on to promote the changes that are needed. Increasing agricultural production requires a conjunction of favorable measures and circumstances; the essence is a mutually reinforcing package deal.

References

Balassa, Bela, and associates. 1971. *The Structure of Protection in Developing Countries.* Baltimore: John Hopkins University Press.

Binswanger, Hans P., and Mark R. Rosenzweig. 1986. "Behavioural and Material Determinants of Production Relations in Agriculture." *Journal of Development Studies* 22, no. 3: 503–39.

Bromley, Daniel W. 1981. "A Discussion of the Role of Land Reform in Economic Development: Policies and Politics." *American Journal of Agricultural Economics* 63: 399–400.

Clapham, Christopher. 1985. *Third World Politics: An Introduction.* London: Croom Helm.

Faaland, Just, ed. 1982. *Population and the World Economy in the 21st Century.* Oxford: Basil Blackwell.

Gaitskell, A. 1959. *Gezira.* London: Faber and Faber.

Gower, David D., and Jerry Vansent. 1983. "Beyond the Rhetoric of Rural Development Participation: How Can It Be Done?" *World Development* 11, no. 5: 427–46.

Griffin, Keith. 1979. *The Political Economy of Agrarian Change.* 2d ed. London: Macmillan.

Hirschman, A. O. 1958. *The Strategy of Economic Development,* New Haven: Yale University Press.

Horton, Douglas. 1986. "Assessing the Impact of International Agricultural Research and Development Programs." *World Development* 14, no. 4: 453–68.

Huntington, Samuel P. 1968. *Political Order in Changing Societies.* New Haven: Yale University Press.

Hydén, Göran. 1988. "State and Nation under Stress." *Forum for utviklingsstudier,* no. 6–7.

Janvry, Alain de. 1981. "The Role of Land Reform in Economic Development: Policies and Politics." *American Journal of Agricultural Economics* 63: 384–92.

Khan, A. R. 1979. "The Comilla Model and the Integrated Rural Development Programme of Bangladesh: An Experiment in Cooperative Capitalism." *World Development* 7: 397–422.

Kirk, Colin. 1987. "Contracting Out: Plantations, Smallholders and Transnational Enterprise." *IDS Bulletin* 18, no. 2: 45–51.

Maier, Karl. 1989. "Mozambique's Ruling Party Drops Marxism." *Independent,* London, 31 July.

Myrdal, Gunnar. 1968. *Asian Drama.* New York: Twentieth Century Fund.

Nove, Alec. 1988. *An Economic History of the U.S.S.R.* Harmondsworth: Pelican Books.

Parkinson, J. R. 1965. *Agricultural Co-operation in Northern Ireland.* Cmd. 284. Belfast: Her Majesty's Stationery Office.

Pearse, Andrew. 1980. *Seeds of Plenty, Seeds of Want.* Oxford: Oxford University Press.

Revelle, Roger. 1982. "Resources." In Just Faaland, ed., *Population and the World Economy in the 21st Century,* pp. 50–77. Oxford: Basil Blackwell.

Ruttan, Vernon W. 1984. "Integrated Rural Development." *World Development* 12, no. 4: 393–401.

Schultz, Theodore W. 1964. *Transforming Traditional Agriculture.* New Haven: Yale University Press.

Sicular, Terry. 1988. "Plan and Market in China's Agricultural Commerce." *Journal of Political Economy* 96: 283–307.

Thomas, John W., and Merilee S. Grindle. 1990. "After the Decision: Implementing Policy Reforms in Developing Countries." *World Development* 18, no. 8: 1163–81.

Timmer, C. Peter, Walter P. Falcon, and Scott R. Pearson. 1983. *Food Policy Analysis.* Baltimore: Johns Hopkins University Press for the World Bank.

World Bank. 1986. *World Development Report, 1986.* New York: Oxford University Press for the World Bank.

——. 1988. *World Development Report, 1988.* New York: Oxford University Press for the World Bank.

I I

Notes on Agriculture
and the State

Raymond F. Hopkins

State intervention in agriculture has a long tradition. Lindert presents evidence in Chapter 2 for a changing pattern of agricultural policies in the course of economic growth. In early modern European and contemporary developing countries, the state taxes agriculture; in modern industrial states, the government subsidizes agriculture. This pattern does not arise from economic rationality (usually) but from political economic forces ascendent at a particular time in a nation's history.

Political economy does help explain this evolution of agricultural policy. The dynamics are more complex, however, than those usually elaborated by economists. The purposes and consequences of state action are often divergent. Consequences are frequently unintended and sometimes perverse. Moreover, the very evolution of the state is closely linked to the development of agriculture and the effects that agricultural policies have upon it.

Anthropologists have closely linked the expansion of governing institutions—from minimalist governing systems to complex, modern state systems—with changes in agricultural production. The need to regulate market activity and resolve land disputes for settled agriculturalists, for instance, is postulated as the basis for the rise of African feudal-type systems (Fortes and Evans-Pritchard, 1940; Mair, 1962, pp. 29–31). Likewise, the centralization of state power and national policies in the modern era is linked to changes in agriculture (Barraclough, 1976; Cochrane, 1979). State financing for agricultural modernization and the expansion of markets, some argue, played the critical role in the modernization of Europe and the expansion of the European state system (Wallerstein, 1974; Tilly, 1975; Tracy, 1982). The agricultural transformation is now seen as essential for successful economic development in late-developing countries, such as Turkey, and very late developers, such as Sub-Saharan African countries (Johnston and Kilby, 1975; Eicher and Staatz, 1984; Mellor, Delgado, and Blackie, 1987; Akay, 1988). These countries typically have the largest portion of their labor force in agriculture and rely on agricultural export earnings for developmental capital goods imports. Their populations spend 40 to 70 percent of their household income on food.

Agriculture is regarded as central to developing an economy and to enhancing or undermining state authority (Scott, 1975, 1985; Hopkins, 1986, 1988). Given this centrality, two key questions arise. Why have states intervened in agriculture in the particular ways they have? Why have some interventions been successful, while others have been failures?

Purposes and Consequences

In different historical situations, states formulate different agricultural policies. They differ because they seek particular goals and choose varying instruments of policy. It is difficult, of course, to be certain about the real purposes of states, as opposed to the merely stated ones. Furthermore, though formal, stated policies are generally accessible to the historian or economist in the form of explicit legal actions and recorded state expenditures, the actual implementation of policies and the use of state funds may vary considerably from those formally stated. This is particularly true in societies in which state capacity is weak, or a "soft" state, to use Myrdal's term.[1] Throughout contemporary Sub-Saharan Africa, for instance, where countries had a dismal record in agricultural performance in the 1970s and 1980s, an examination of policies for regulating markets, subsidizing agricultural inputs, fixing prices, and even creating nutritional safety nets exposed a wide gap between official policy and actual performance. Effective policy instruments may thus be highly

[1] This idea of "soft states" is discussed in Chapter 10 by Faaland and Parkinson, using Myrdal's 1953 distinction for Asian states.

limited in periods of nascent state formation. Although rural populations often lack organization and appear vulnerable to the interests of the powerful, they nonetheless may pose a formidable obstacle to state manipulation, whether in Africa in the late twentieth century or in Europe in the eighteenth and nineteenth centuries (Chapter 10 by Faaland and Parkinson; Hydén, 1980; Scott, 1985; Cohen, 1988; Glickman, 1988).

These distinctions allow us to consider the historically dynamic process of the role of the state in agriculture. First, purposes for state action are conceived. Second, policy choices are made. The third stage is implementation, when the tools chosen are used. Finally, consequences occur that in turn affect initial purposes. This dynamic is pervasive in the history of relations between the state and producers, merchants, and consumers. On the one hand, cases exist in which the state has used its resources to promote efficient agriculture, for example, through provision of collective goods, with results that are positive for both economic and noneconomic values.[2] Several Asian states are such cases. On the other hand, states can exploit agriculture, thus undermining growth opportunities and alienating segments of society, as the research by Valdés in Chapter 3 suggests.

Political Economy Considerations in the Evolution of Agricultural Policy

The emergence of higher standards of living, thanks to industrialization, has led to a welfare role for states. Governments cannot regulate markets solely in the interest of efficiency and social profitability (if they ever could); they also must redistribute social values to ensure some degree of equity or justice (Okun, 1975). In the last several centuries, responsibility for administering to the needs of weak and vulnerable people has shifted from the private to the public

---

[2]For purposes here, the outcomes of state agricultural transactions are referred to both in terms of benefits that are directly economic—in the sense that they yield monetized effects whose net benefits and costs can theoretically be assessed using standard economic accounting methodologies—and in terms of noneconomic benefits, which include important aspects of human behavior, such as loyalty to the government, voluntary compliance with policy, national self-esteem and rectitude, and other values. These are not monetized directly. Even shadow prices for such values would be hard to calculate since their manifestations in society often occur in step-level events. Changes in such values, however, are conceptually discrete movements. For example, government legitimacy can vary by degrees, but changes in government legitimacy outside revolutionary situations are not readily measurable.

sector. The welfare state, with its plethora of programs that provide citizens minimum guarantees of goods and services, is a manifest result of this shift.

Food price policy has been a particularly important instrument in developing countries affected by this shift in public demands. To guarantee access to basic foodstuffs and augment the household income of the extreme poor, many policy makers have adopted such measures as fixed prices or subsidy policies. Ethical considerations, arising from the very fabric of society itself, lead to this redistribution on behalf of the poor (Chambers, 1983). Most recently, this concern of the government for the equity and the poor has been manifested (even in very poor states) in such initiatives as the UNICEF proposal for "adjustment with a human face" and the World Bank's effort to achieve food security in Africa.[3] Such policies are not without economic costs, however; the frequent trade-offs between equity and efficiency, between short- and long-term consequences, and between economic and non-economic values, become especially poignant in cases when government capacity is already constrained by slow rates of economic development.

Models of Political Economy

Economists frequently criticize government policy that distorts markets. They argue that such interventions lead to non-Pareto optimal outcomes, reduce efficiency, slow the expansion of the production frontier, promote disincentives, and protect the unduly privileged. Such criticisms arise not only from neoclassical assumptions from which most economists approach social analysis but also from a genuine concern to seek better mixes of purposes and outcomes from government intervention. Market failures, exploitative government behavior, and policies encouraging stagnation rather than economic growth seem pathological from this perspective. To account for such policy failures, economists frequently blame "poli-

[3]In Ghana, for example, the government adopted in 1987 a Programme of Actions to Mitigate the Social Costs of Adjustment (PAMSCAD), which was designed to offset the disproportionate or unfair burden of adjustment on vulnerable groups in Ghanaian society brought about by the economic recovery program initiated in 1983. PAMSCAD is supported by the World Bank, the International Monetary Fund, and bilateral donors. This intervention was seen as a short-term policy designed to cushion the effects of adjustment that were proving excessively harmful to the most vulnerable. Nevertheless, its total cost is estimated to be U.S. $84 million, which represents over 10 percent of annual government revenues.

Figure 11.1. Three models of political economy

Model I: State as Arena

Groups outside government, based on rational interest calculations, seek to influence policy.

Variant A:
Competitive, pluralist system: multiple groups, with changing alliances. Failure of public interest arises from divisible benefits that provide incentives to some groups (farmers) to pressure for policy preference, while more diffuse, larger groups are less active because problems of collective benefits offer weak incentives to mobilize.

Variant B:
Noncompetitive, class-dominated situation: a group largely external to state officials (e.g., bourgeoisie, salariat, ethnic groups) dictates policy.

Model II: State as Actor

State officials and titleholders act to maximize their values (wealth, safety, affection, and so on). If the state has a high discount value, its leadership usually self-destructs. Its features are rent seeking, bureaucratic self-protection and accumulation, and extortion by individuals. The state is seen by itself and others as competing with society to maintain the privileges of the state officeholders. If the state has a low discount rate, leaders may move toward a broader incorporation of popular interests with state interest—a transformation, especially in "weak" states, toward a Model III type.

Model III: State as Builder

Goals are hierarchical to meet sovereign nation-state desiderata: security, growth, and welfare. Weak states, typical among LDCs, give high priority to inculcating habits of compliance and improving the probability of enforcement. Security, particularly domestic, is a central issue. As the state as agent becomes stronger, its capacity and interest in serving national goals move it to allocate more resources or allow more risk in policies aimed at economic growth and, eventually, welfare. Weak states that prematurely give high priority to economic growth and welfare frequently fail.

There are several key questions. Which mix best characterizes an actual state at a particular time? For that context, what advice about economic policy is most appropriate? For policy advisers who usually share Model III goals, what policies best meet the preference of the state?

---

tics." More recently, an analysis based on political economy has sought to interpret the development and change of policy in various historical contests (Staniland, 1985; Bates, 1989). Three basic approaches to political economy can be outlined (see Figure 11.1).[4]

State as Arena. In the first model of political economy, the state is an arena for competing interest groups. Model I is the most prevalent political model for describing the basis of government action,

[4]Each of the political economy models may be found in contemporary analyses of various writers.

particularly among economists (Anderson, 1987; Lele and Christiansen, 1989). Powerful interests, often urban based, as in a white-collar salariat class, or grounded in powerful landowners have partial or complete control over the instruments of the state and use them to advance their own interests.[5] For these powerful groups, the only trade-off is between short- and long-term gain; otherwise they promote their group's rational choice strategies for state action, which will only coincidentally promote the interests of the society as a whole (Bates, 1981).

State as Actor. In the second model of political economy, the state is an actor in its own right. The clan of tribal societies, the royal families of the feudal ages, and the modern bureaucratic state with its cadres of officials are examples of the state as a rational calculator of costs and benefits for maximizing state power and the income of its officials. Such calculations are, of course, constrained by the pliability of the state's subjects and the technology the state can use to enforce its will as well as to foster economic efficiency. The basic calculus, however, derives from the interests of those running the state, whether royalty, a privileged class, or an entrenched bureaucracy.

Examples of the state as self-interested actor range from the reign of France's Louis XIV with his diffidence toward those outside his state (*"l'état c'est moi!"*) to the kleptocracy of Zaire.[6] Activities of rent-seeking states have been caustically described by citizenries of countries ranging across the ideological spectrum. The Soviet Union, under Glasnost, has printed numerous complaints about management both in agriculture and in officialdom generally. Such popular complaints about state-controlled exploitation are widely reported in the literature on dependency in Latin America and Africa; attitudes with the same valence are voiced more gently in criticisms of the heavy hand of government expressed by the American farm population (Cochrane, 1979).

State as Builder. In the third model of political economy, the state focuses on building its capacity. Weak states seek power but not as an end in itself, as in the case of Model II. The state attempts to build support and discover policies that will best serve purposes re-

[5]For a discussion of urban bias, see Lipton (1977).
[6]For Zaire, see Callaghy (1984).

quired for survival of national sovereignty (Krasner, 1988). According to this model, the key distinction among states and their policies arises from the state's capacity—as ranked on a continuum from a weak (or soft) state to a strong (or hard) one—to be an agent that is either ephemeral and elusive or tough and effective. The writ of state authority itself is the issue in question (Huntington, 1968). Often the state's capacity extends no further than the capital city or the personal friendships of top leaders. A distinction between formal, de jure but ineffective states and stronger de facto states is particularly apt in modern conditions in Africa. Since 1980 several writers have alluded to the inability of the state to adopt policies that genuinely regulate the economy—that demonstrate capacity beyond control over imports and exports. Even in this realm, smuggling can be a major element allowing agriculture to escape state regulation (Hydén, 1980; Bratton, 1989). It is important to recognize that even in countries where states represent a powerful element in society, such as the United States and the Soviet Union, state policy is not solely in control. Other factors, particularly implementation problems and reactions of individual producers or consumers, frequently lead to policy outcomes quite different from those expected or predicted by sophisticated analysis. In these conditions, actions by the state to intervene in agriculture, whether to support producers or consumers or to stimulate and redistribute wealth, may also represent a series of trials and errors in policy formation.

Purposes Served by State Intervention in Agriculture

In analyzing the history of agricultural policy in various countries, it may be useful to identify which of the three models of the state best fits the evolution of policies, either over all cases or at particular times, and the purposes served by state intervention. States as actors for themselves (Model II) are aggrandizing in character but frequently have short-term successes and long-term failures. Perhaps the Philippines under Marcos fits this pattern. A state acting as an arena for competing groups may become captured by narrow interests, whether of powerful landlords or military officers, which may lead to important policy distortions and to lost opportunities for the economy as well as disaffection of the population (Huntington, 1968). The third vision, the state as would-be entrepreneur, may best account for states that intervene in society primarily to bring order and some semblance of control over agriculture; the purposes of such weak states may be highly transient. Since the state is not deeply institutionalized—its leadership and circumstances change fairly quickly—it might also be opportunistic. The state's search for optimizing behavior takes place under high uncertainty. Ironically, because weak states are not anchored in tradition or legal formalities, they may be more erratic in the policies they follow, but they are also more influenced by policy advice given by economists. Policies of developing countries, especially those in Africa, frequently fit this model. The use of agricultural policy to advance development might be most critical in weak states; they have the greatest opportunities to restructure agriculture, particularly after a revolution or foreign conquest—for example, the land reforms in Japan, Taiwan, or China.

Six purposes seem to explain historical evolution in agricultural policy. Government intervention, whether best understood from the viewpoint of Models I, II, or III, has nearly always involved some mixture of these purposes, which are outlined below. Consequences of such efforts have also shaped future capacity for undertaking policies. Shortsighted, unsuccessful interventions can harm both the state and agriculture. The success of the economic transformation of agriculture, and the economy more generally, and the development of national loyalties and institutionalized state structures are all deeply interrelated consequences of evolving agricultural policy.

Extract Resources from Agriculture. The first purpose, classic for self-serving or rent-seeking states (Model II), is to extract resources from the agricultural sector for the purpose of state maintenance, including guaranteeing a high standard of living among official or royal classes. Since such extraction from the production or exchange of agricultural products serves only to redistribute wealth to office-holders and central state authorities, it represents the purest case of exploitation. Such action is the functional equivalent of mafioso-style extortion in the private sector. The government's treatment of French peasants before the 1789 revolution is a classic instance of such a purpose dominating state policy. Zaire in the 1970s is another instance (Callaghy, 1984; Scott, 1985). In Models I or III, extracting resources from agriculture may be linked by expenditure policies to more altruistic intentions and even consequences.

Expansion of the State. The state intervenes to expand its connections throughout society. The expansion of the state, for good or ill, requires replacing local fiefdoms and baronies with the imprint of central authority. States thus devise policies that require low investment in personnel and offer an opportunity to present central authority as a positive force in the life of the peasantry (Laski, 1938; Moore, 1967; Bratton, 1989). Capitalist agriculture, for example, required centralized authority over local manor systems or tribal economies; the substitution of state regulations for such systems made possible the encouragement of capitalist practices. The state acted to assert its authority, however, rather than to base its policies on a theory of economic development. This assertion of authority was most often the core purpose for such action (Tilly, 1975).

Protect Agriculture as a Resource. At times the state has intervened to put agriculture on a competitive basis with other economic sectors. By nature, agriculture is a risky business. Climatic forces make crop yields uncertain. Protection of land tenure rights and fair marketing arrangements for the often poor and disorganized farmers depends on laws and government. Producers who provide the physical labor in agriculture, as distinguished from large landowners and managers, frequently have little power over the affairs of state. Such numerous but disorganized elements of society lack the free time or direct rewards to organize and pay the cost of collective bargaining with the state (Dahl, 1962, pp. 55–71; Olson, 1965; Lindblom, 1977). The state can serve to secure socially efficient collective benefits, which the "free rider" problem would otherwise cause to be neglected.

Promote Economic Development. The state undertakes various measures to stimulate economic development, such as investment in agricultural research, encouragement of new technology, or greater guarantees of profitability to producers taking risks or investing more of their own labor. This role of the state is the classic one assumed by most economists (given the normative assumptions within which most of their work is cast). With this purpose in mind, analysts carefully try to assess the optimal benefit-cost ratios of various government investments to maximize efficiency among producers, lower marketing costs, and alleviate uneconomical fluctuations in demand and unemployment among the poor.

Improve Welfare of the Poor. The promotion of equity and the meeting of human needs is often cited as a goal of government policy.

Subsidies targeted to the hungry poor, absorption of the adjustment costs for those moving out of agriculture, and other state-funded compensatory actions may not have positive rates of return on investment but are justified by basic ethical considerations and, secondarily, perhaps by the goal of state survival as a national, social instrument (see the purpose of political stability below). Such interventions to assist the poor might be a drag rather than a spur to general economic development. Egypt and Sri Lanka, for example, have been cited as cases in which the burden of food subsidies, equaling 10 to 20 percent of total government revenues in the 1970s, was for economic growth, a long-term negative factor. While industrialized states, such as the United States, Europe, and other members of the Organization for Economic Cooperation and Development, may be able to afford welfare state policies that include targeted food guarantees through programs such as food stamps, institutional feeding, and direct distribution, their costs are modest. Such redistribution, however, weighs heavily on states with lower incomes, less efficient economies, and a large portion of the population employed in agriculture (Pinstrup-Andersen, 1988).

Promote Political Stability. Even from a strict economic perspective, political stability, however difficult to estimate, is worth some economic benefit.[7] Maintenance of political authority reflects, in part, the political ties, both personal and ideological, between state leadership and the rural sector of the economy. Thomas Jefferson regarded the agricultural ethic as the basis of American democracy—an argument that has been supported two centuries later (McConnell, 1952; Hadwiger and Talbot, 1979). Analogously, agriculture as an embodiment of state virtue has flourished under Félix Houphouët-Boigny of Côte d'Ivoire. Houphouët-Boigny proclaims himself the country's "number one peasant." Emotional and affective ties, therefore, can bind agriculture and the state in ways that sustain national character, project cultural values, and bolster political stability. These cultural forces can emotionally distort the rational choice template often placed upon government intervention (Potter, 1954; Hadwiger and Talbot, 1979).

In summary, the state intervenes in agriculture usually to accomplish one or several of these six purposes. Its success or failure fre-

---

[7]In Chapter 5, for example, Timmer cites the need for government intervention to accelerate efficient growth and income gains to farmers and access to food for low-income consumers, but he also argues that these interventions and others are important to maintain political legitimacy.

quently depends on political economic factors. In Model I the state is an arena for powerful private forces to fix public policy. In Model II the state is a maximizer on behalf of itself as an actor—that is, it maximizes the private interests of officialdom. In Model III the state is also an actor, but one motivated by sovereignty goals and highly limited by missing information, uncertain popular loyalties, and ineffectual instruments. The three models are not mutually exclusive; they do, however, organize distinctive analytical elements to explain the actions of a state. In any actual case, some mixture of all three models is likely, but in most cases, one or another model will prove more illuminating and predictive of state action than others, especially with respect to the state's purpose and effect in interventions in agriculture.

References

Akay, A. Adnan. 1988. "From Landlordism to Capitalism in Turkish Agriculture." Working Paper 12. Milton Keynes, U.K.: Open University.

Anderson, Dennis. 1987. *The Public Revenue and Economic Policy in African Countries*. Discussion Paper 19. Washington, D.C.: World Bank.

Barraclough, Geoffrey. 1976. *The Crucible of Europe*. Berkeley: University of California Press.

Bates, Robert. 1981. *Markets and States in Tropical Africa*. Berkeley: University of California Press.

———. 1989. *Beyond the Miracle of the Market*. New York: Cambridge University Press.

Bratton, Michael. 1989. "Beyond the State." *World Politics* 41 (April): 407–30.

Callaghy, Thomas. 1984. *The State-Society Struggle: Zaire in Comparative Perspective*. New York: Columbia University Press.

Chambers, Robert. 1983. *Rural Development: Putting the Last First*. London: Longman.

Cochrane, Willard. 1979. *The Development of American Agriculture: A Historical Analysis*. Minneapolis: University of Minnesota Press.

Cohen, Ronald, ed. 1988. *Satisfying Africa's Food Needs*. Boulder, Colo.: Lynn Reinner Publishers.

Dahl, Robert. 1962. *Modern Political Analysis*. Englewood Cliffs, N.J.: Prentice-Hall.

Eicher, Carl K., and John M. Staatz, eds. 1984. *Agricultural Development in the Third World*. Baltimore: Johns Hopkins University Press.

Fortes, M., and E. E. Evans-Pritchard, eds. 1940. *African Political Systems*. London: Oxford University Press.

Glickman, Harvey, ed. 1988. *The Crisis and Challenge of African Development*. New York: Greenwood Press.

Hadwiger, Donald F., and Ross Talbot. 1979. "The United States: A Unique Development Model." In Raymond F. Hopkins, Donald J. Puchala, and Ross B. Talbot, eds., *Food, Politics, and Agricultural Development*, pp. 21–44. Boulder, Colo.: Westview Press.

Hopkins, Raymond F. 1986. "Food Security, Policy Options and the Evolution of State Responsibility." In F. Lamond Tullis and W. Ladd Hollis, eds. *Food, the State, and International Political Economy*, pp. 3–36. Lincoln: University of Nebraska Press.

———. 1988. "Political Calculations in Food Subsidies." In Per Pinstrup-Andersen, ed., *Food Subsidies in Developing Countries*, pp. 107–26. Baltimore: Johns Hopkins University Press.

Huntington, Samuel. 1968. *Political Order in Changing Societies*. New Haven: Yale University Press.

Hydén, Gören. 1980. *Ujamaa*. Berkeley: University of California Press.

Johnston, Bruce F., and Peter Kilby. 1975. *Agriculture and Structural Transformation*. New York: Oxford University Press.

Krasner, Stephen. 1988. "Sovereignty: An Institutional Perspective," *Comparative Political Studies* 21 (April): 66–94.

Laski, Harold J. 1938. *The Rise of European Liberations*. London: Unsrin Books.

Lele, Uma, and Robert E. Christiansen. 1989. "Markets, Marketing Boards and Cooperatives: Issues in Adjustment Policy." Washington, D.C.: World Bank, MADIA Project. Typescript.

Lindblom, Charles E. 1977. *Politics and Markets*. New York: Basic Books.

Lipton, Michael. 1977. *Why Poor People Stay Poor: A Study of Urban Bias in World Development*. London: Temple-Smith.

McConnell, Grant. 1952. *The Decline of Agrarian Democracy*. Berkeley: University of California Press.

Mair, Lucy. 1962. *Primitive Government*. Baltimore: Penguin.

Mellor, John W., Christopher L. Delgado, and Malcolm J. Blackie, eds. 1987. *Accelerating Food Production in Sub-Saharan Africa*. Baltimore: Johns Hopkins University Press.

Moore, Barrington, Jr. 1967. *Social Origins of Dictatorship and Democracy: Lord and Peasant in the Making of the Modern World*. Boston: Beacon Press.

Okun, Arthur. 1975. *Equality and Efficiency: The Big Tradeoff*. Washington, D.C.: Brookings Institution.

Olson, Mancur. 1965. *The Logic of Collective Action*. Cambridge: Harvard University Press.

Pinstrup-Andersen, Per, ed. 1988. *Food Subsidies in Developing Countries*. Baltimore: Johns Hopkins University Press.

Potter, David. 1954. *People of Plenty*. Chicago: University of Chicago Press.

Scott, James C. 1975. "Exploitation in Rural Class Relations." *Comparative Politics* 1 (July): 489–532.

———. 1976. *The Moral Economy of the Peasant*. New Haven: Yale University Press.

———. 1985. *Weapons of the Weak*. New Haven: Yale University Press.

Staniland, Martin. 1985. *What Is Political Economy?* New Haven: Yale University Press.

Tilly, Charles, ed. 1975. *The Formation of National States in Western Europe*. Princeton: Princeton University Press.

Tracy, Michael. 1982. *Agriculture in Western Europe: Crisis and Adaptation since 1880*. 2d ed. London: Jonathan Cape.

Wallerstein, Immanuel. 1974. *The Modern World System: Capitalist Agriculture and the Origins of the European World Economy in the Sixteenth Century*. New York: Academic Press.

12

What Have We Learned?

C. Peter Timmer

The 1990s will be a challenging decade for economists working to reduce poverty and speed agricultural development. Not only were the 1980s a "lost decade" for many countries of the Third World, especially in Africa and Latin America, but the turn of the decade brought Eastern Europe, the Soviet Union, and much of socialist Asia back on the agenda of development economists. Suddenly the questions asked about the role of the state in the development process take on new urgency as the observed range of economic structures and interventions widens dramatically.

Many countries newly independent after World War II sought their models for development strategies in the Soviet example, with its massive displacement of market forces by central planning and the decimation of agriculture in support of forced-pace industrialization and an urban proletariat. Eastern European countries had the model imposed on them by the Soviet Union. Four decades later, the legacy of Stalinist economics is not just the ruined economies of Poland, Czechoslovakia, and the rest of Eastern Europe and the Soviet Union, with their outdated industrial sectors, backward agricultures, and near vacuum of market institutions. The legacy extends to much of the Third World as well, but there the response to the political revolutions that swept Eastern Europe and the Soviet Union at the end of 1989 has been much more muted. There is near unanimity in Eastern Europe that casting off the Stalinist structure as quickly as possible is imperative and that the only feasible way to do it is by introducing unrestrained "Dickensian" capitalism. The state can be trusted with nothing but defending the border.

Such an approach would have been unthinkable at the end of the 1970s and is still strongly challenged in the Third World. But the intellectual revolution that transformed tax codes in rich countries and privatized state-owned enterprises in poor ones has provided the foundations for an entirely new balance between the public and private sectors. Perhaps the most perplexing question now is how far the pendulum must swing against state intervention on behalf of social goals before there is a return to the mixed economy that still prevails in the United States, Western Europe, and the highly successful societies of East Asia. Must countries live with the abuses and inequities of full-blown capitalism in order to get rich? Indeed, will a hands-off approach to free markets even allow poor countries to get rich? Can we identify the nature of those states that can be trusted to intervene on behalf of economic growth and social welfare and be able also to spot the incipient venal or corrupt state? If so, what pressures can be brought to bear to steer societies in the right direction? Are ideas and articles enough? Can pressures from the donor community be productive? Or will a "global policeman" have to step in and remove the General Noriegas of the world? These are, of course, ancient questions, to which only partial answers can be provided.

The Timing and Scope of Government Interventions

The chapters in this volume are on diverse topics and do not add up to a systematic treatise on the role of the state in agricultural development. Still, there is an underlying consistency to the key themes these chapters pursue, a consistency that should give pause to proponents of free markets and minimal state intervention as the surest path to riches. In any given empirical context, it is difficult to judge whether government interventions or market forces would be most effective in promoting economic development and broad-based increases in material welfare. Whether the issue is price policy, trade strategy, employment schemes, rural development projects, or the management of food aid, the question of when to intervene to alter market-determined outcomes is at the core of both analytical and political debates. Analysis of all these issues is conducted on the premise that adequate market institutions exist for neoclassical models to illuminate the nature of alternative equilibria resulting from different types and degrees of government intervention. This is precisely the premise that does not hold in most of the socialist countries now embarked on structural and political reform. It is one more reason why development economists, with their concern for structure and market failures, may have more insights into how to pursue those reforms than economists who primarily study Western economies.

The nature of the state, including its political legitimacy, ideological orientation, and bureaucratic capacity, provides one crucial element in determining when to intervene. But other elements are also important. For example, the competitive pressures Vietnam feels from successful neighbors—especially Thailand, Taiwan, Malaysia, and Indonesia—have pushed it into restructuring an economy nearly destroyed by ideological commitment to state socialism. The economic reforms in Vietnam far surpass any yet accomplished in Eastern Europe, whereas political reform lies frozen, as in China. Another emerging wisdom of economic reform in Eastern Europe is that only popular democracies can engender the political support needed for the painful readjustments that come with the introduction of untrammeled capitalism. If the economy of Vietnam grows at twice the rate of those in Eastern Europe to the turn of the century, as now seems possible, such wisdom must surely be reconsidered.

Personalities also matter. Leadership qualities and personal values of individual leaders can wreck a rural economy, as in Tanzania, or rescue it from decades of urban bias, as in Indonesia. The revival of interest in the role of monarchies in establishing political unity and social cohesion reflects this realization that nation-building requires more than an independence movement sitting with the reins of government.[1]

Even economic analysis sometimes makes a difference to the scope and nature of government interventions. Calculations showing the efficiency of market-led decisions with respect to choice of technique in rice milling once kept a government from subsidizing "modern" but labor-displacing rice mills. By contrast, the same government was persuaded to overrule free-market forces determining the demand for fertilizer because a large subsidy designed to speed the adoption of this critical yield-raising input was shown to be socially profitable.[2] A valuable lesson in political economy was learned subsequently;

---

[1]See the editorial "The Comeback of Kings" in the April 14, 1990, issue of the *Economist* (p. 18) for a lighthearted but surprisingly serious discussion of this revival.

[2]Both episodes are described in Timmer (1989b).

econometric arguments defending the social profitability of the fertilizer subsidy were far more effective than critiques by the same analysts when economic circumstances no longer justified the large subsidies.

Several authors in this volume have attempted to categorize different "natures" of the state relative to the tasks of agricultural development. The treatment by Hopkins from the perspective of political science in Chapter 11 offers a succinct language for debate, but the operational potential of treating the state as an arena, an actor, or a builder remains elusive. Indeed, as Hopkins notes, the three models of the state characterized by these titles merge subtly one into another. There is considerable debate, for example, whether the South Korean state was primarily an actor or a builder during the growth spurt from the 1960s into the 1980s and whether it now functions as an arena for competing political and economic interests. The answers to such questions are important to the efforts of trading partners in determining the best strategic approach to open up the Korean economy; they are possibly important for internal policy analysts as well.

But the implications for the basic models economists use to evaluate policies are unclear. How far, for example, should social benefit-cost analysis go to incorporate the politically *feasible* into models that are currently designed to illuminate what is socially *desirable*? The answer is surely different depending on whether the state is an arena for competing interest groups, a rent-seeking actor, or a strong builder with a long time horizon. But how should the nature of the state be included? The answers offered so far by political economists are primarily descriptive, seeking to explain past policy actions. Sometimes these efforts are remarkably successful, as with Lindert's analysis in Chapter 2 of biases in agricultural pricing. But any normative appraisal of these results remains firmly rooted in standard neoclassical models of Pareto optimality and efficient resource allocation. It is not clear that these models are robust in a complicated world of political economy.

General Patterns and Particular Circumstances

If market failures and government failures are empirical issues that depend on local circumstances, careful analysis that is alert to both types of failure can help government policy makers in all three models of the state to steer a pragmatic path and avoid crashing on either shore. As noted in Chapter 1, analysts have the guidance of history, comparative experience, and a growing theoretical literature on the political economy of economic development to help organize their work. Certain patterns of success and failure with markets and interventions are quite robust, especially the lesson that displacing markets is far less likely to stimulate rising agricultural productivity and rural growth than strategies that seek to make markets more effective and competitive.

By contrast, some patterns of intervention are widespread despite widely varying circumstances. Protection for producers of agricultural commodities facing rising competition from rapidly growing industrial sectors is pervasive, despite the hostility of economists to the severe distortions created. Basic political forces, quite possibly with an underlying economic logic that is too subtle for economists to model or measure, must explain such powerful pressures on policy makers in democratic and authoritarian governments alike. Similarly, industrial protection as a stimulus to import substitution discriminates against agriculture in every economy in which it is implemented; so too must very basic mechanisms of general equilibrium force this result in both capitalist and socialist economies. Even without the linkages imposed by functioning factor and product markets, industrial autarky imposes heavy burdens on agriculture and ultimately on the overall process of economic growth. By contrast, it is hard to name poor countries that have been negatively affected by serious efforts to stimulate agricultural growth (just as it is easy to name rich countries that have).

The Desire for Both Equity and Growth

Perhaps the most powerful lesson from the 1980s is the importance of economic growth to solve what are otherwise considered to be distributional issues. Without rapid economic growth, no economy has been able to sustain an egalitarian distribution of food. Redistribution *without* growth has powerful and negative incentive effects on the rural economy; the only way to reduce poverty and hunger is to raise labor productivity, employment, and real wages of unskilled labor. As with the general-equilibrium effects of industrial protection, intersectoral linkages across labor markets are sufficiently strong that any unisectoral approach to raising employment or wages is doomed to failure. Designers of rural development projects have learned this lesson. Although individual projects might have enough human and financial resources poured into them to serve as "successful" showcases, replication requires a favorable economic and institutional environment.

Staying the Course. Such an environment is fragile, however, and shortsighted state interventions are the most common threats to the delicate institutions, investors' expectations, and community self-confidence that are essential to the successful spread of rural development projects in market-based economies. There are reasons, of course, for the short time horizon of so many governments, extending to the overarching concerns for political legitimacy, maintenance of power, and seeking opportunities for rents. Concerns for food security, however, especially in urban areas, are often the basis for many myopic actions that have devastating long-run consequences for the development of market institutions. Governments do foolish things in the name of stabilizing food prices, including commandeering food at gunpoint, banning private stockholding, and enacting legal price ceilings without the logistical capacity to make them effective. In the name of maintaining access of the poor to basic staples in the present, such interventions are one of the surest ways to deprive them of that access in the future.

Food Aid to Protect the Welfare of the Poor. This short time horizon on the part of food planners in developing countries is matched by that of donors. Despite the rhetoric to the contrary, there is little evidence to suggest, for example, that food aid can be used as a reliable source of supplies for planning food security or that food aid resources can be used effectively in agricultural development. Part of the problem is a familiar Catch 22. Only countries with sound food policies and the analytical capacity to program food aid supplies into an efficient food system can use these resources effectively. But such countries do not need food aid by the criteria of most donors. By contrast, countries in dire need usually have ineffective food policies that impede the development of the agricultural sector. The dumping of food aid supplies in such environments compounds the problems of development.

This limited potential for food aid is surprising, particularly in view of the vast political resources used to defend its funding and the substantial financial and bureaucratic resources needed to administer food aid programs, which have high opportunity costs in both donor and recipient countries. It might be tempting to conclude that food aid has such a poor record because it inherently involves the state in some form of intervention, either directly in markets or indirectly through more targeted distribution programs. Alternatively, far more of the onus might be put on donor governments for their inability to program and deliver supplies with adequate lead time for long-term planning of food security programs or with sufficient flexibility in the short run to cope with fluctuating needs as domestic supplies fall short. This pessimistic assessment, however, is not warranted for emergency food aid for famine relief. Here the record shows that the donor community has relatively fast response times, and the major bottlenecks tend to be at the recipient end, either for political reasons or because of poor infrastructure and unresponsive bureaucracies.

It is easy to conclude that food aid is irrelevant to the development process. Such a conclusion misses two potential contributions. First, discussions about food aid are often the only effective policy dialogue that donors and countries have about agricultural issues. In a surprising number of countries, only during food aid negotiations do they undertake a careful review of their agricultural and rural development strategies. Food aid does not always lead in this direction, as continuing high levels shipped to Egypt and El Salvador indicate, but the policy potential of food aid negotiations is often quite real.

Second, broader analysis of functional relationships between the state of the world food economy and volumes of food aid actually shipped can highlight important areas in which the world economy fails to serve the interests of the poorest countries. The inverse relationship cited by Falcon in Chapter 9 between world grain stocks and food aid supplies probably runs counter to the needs of these countries. Better mechanisms are needed to allocate food aid, especially ones that would be less perverse in relating needs to availability, and to lessen the instability of prices in world markets directly, thus severing the relationship altogether.

Where Models and History Diverge

Real progress has been made since the world food crisis in the mid-1970s in understanding the role of the state in stimulating or inhibiting agricultural development. Part of that understanding has come by focusing research attention directly on the determinants of state intervention. Development economists now have an acute awareness of the role of interest groups, bureaucratic politics, rent-seeking, and nation-building in the adoption of specific policy proposals.

The underlying lessons from the chapters in this book are neither revolutionary nor even highly controversial. To the rather pragmatic community of development specialists, there are no secrets here. Agricultural development requires a sustained commitment from governments and the private sector, with reasonably clear ground rules on appropriate roles for each. The balance of roles can be surprisingly varied, or at least so the historical record indicates. The lessons on the role of the state summarized above provide only loose guidelines, and it is instructive to speculate on the missing elements in the analysis and the reasons for their absence.

The Dynamic Impact of Investment

Two decades of research and analysis have documented the robust patterns of government intervention in agricultural pricing. Even if economists cannot prevent the distortions, their analysis is more informed and persuasive by knowing why the policies are put in place. And there may even be more economic rationale behind the underlying political motivation, which reflects a desire for price stability as much as it does for pure farmer protection, than existing static models of efficient resource allocation can identify.

But what do we know of historical patterns of investment in the agricultural sector—those on government account and by the private sector, including by farmers themselves? Despite studies of resource flows from rural to urban areas and an entire body of literature on dual economy models of development, which use agricultural surpluses to finance industrial expansion, the answer remains "shockingly little." There is no comparable study of investment flows to and from agriculture to match Lindert's historical and cross-section analysis of pricing biases. The literature on urban bias, initiated by Lipton (1977), suggests that the misallocation of investment funds on public account is massive. And yet no systematic and quantitative assessment has been conducted because the basic data do not exist. Even for public expenditures, it is difficult to identify shares that go to agriculture, to rural areas (not the same thing), and to improving the access of urban markets to rural products (where presumably both ends of the market chain will benefit). Not a single country in the Third World has a time series of private investment in agriculture that includes the value of on-farm savings from both financial and sweat equity.

It is no wonder that the profession remains largely ignorant of the determinants of agricultural investment, in sharp contrast to the progress that has been made in understanding pricing policy. This ignorance creates immediate problems for any attempt to model the dynamic impact of government interventions on the agricultural sector directly and, by extension, on the growth process. For example, the impact of price changes will be captured only as short-run changes in supply and demand. In the absence of functional information on rural investments, models cannot accurately capture indirect effects on savings in rural households. In particular, investments at the farm level in land productivity, livestock, and tree crops cannot be distinguished from investments in financial savings, human capital, or portfolio diversification in the form of support for an urban migrant. Such functions are impossible to specify in the absence of empirical information on the underlying flows, which is precisely what is missing. Only detailed historical analyses of individual countries, or even provincial and regional experiences, can supply such information. Without this information and the functional relationships that could be derived from it, analysts can be properly skeptical of any models claiming to describe the dynamic impact of government interventions on the agricultural sector.

An Agricultural Trade Strategy

A similar range of problems with respect to missing data and unasked questions troubles the debate over export-led growth, especially for agricultural commodities. The evidence presented in Chapter 3 by Valdés is very clear (and consistent with Lindert's analysis in Chapter 2); most countries have systematically discriminated against their agricultural sectors through direct and indirect policy instruments. Part of this discrimination, and indeed a substantial part of urban bias, must stem from deep-seated expectations about adverse movements in the agricultural terms of trade in the face of rapid expansion in production and exports. There is no doubt that the aggregate, short-run price elasticity of demand for most agricultural products is substantially less than one (in absolute terms) and is smaller than the equivalent figure for industrial products. Even if the marginal productivity of investments in agriculture is substantially higher than that in industry, such productivity might sit on a razor's edge—even modest increases in agricultural funding would drive its marginal productivity well below the opportunities in industry. It is best, the argument runs, not to expand agricultural output "too fast."

Such an argument does not explain why so many countries have distorted agricultural incentives so massively, leading to sharp losses in market share for export commodities and to poor growth domestically. Other elements in the political economy of these countries must answer this dilemma. These policy failures in some countries, however, especially in Sub-Saharan Africa, opened market opportunities for other countries, especially in Southeast Asia. Would Malaysia and Indonesia have expanded palm oil exports so rapidly if Nigeria and Zaire had also been aggressively expanding their export volumes? Export-led growth based on the agricultural sector has been successful in only a handful of countries. Is this because there is limited wisdom among countries with respect to growth strategies or because of deeper limitations on the "market" for such strategies?

The dynamics of competition for market share play a key role in the "new international trade theory," which is providing very useful insights into potential welfare gains to individual countries that practice "trade strategy" instead of free trade.[3] A trade strategy requires government intervention in precisely those economic activities for which the historical record is most dubious—protection of infant industries, subsidies to exporters, and bilateral quota deals. These measures open obvious opportunities for rent seeking and indirect, but powerful, biases against the agricultural sector. Even leaders in the theoretical analysis of trade strategies, such as Krugman (1987) and Baldwin (1988), are doubtful of the ability of most countries to implement the strategy effectively.

Can such trade strategies be used in the future to stimulate agricultural exports and thus contribute to more rapid economic growth? Many are doubtful about the implementation capacity in most developing countries and wary of the clear potential for subverting the theory of public benefits into private gains. With a global effort to increase agricultural exports, the consistency of price assumptions must also be questioned. Even if diversification of export crops is a technical possibility, the new export pessimism discussed by Valdés in Chapter 3 cannot be dismissed. The marketing infrastructure and commercial skills needed to diversify agricultural exports are available in only a handful of middle-income countries; they are especially lacking in just those Sub-Saharan African countries most vulnerable to price declines for their main agricultural exports. For these countries, a more likely prospect is for "immiserizing growth,"

[3] A very useful summary of this literature appears in Rodrik (1988).

as prices fall faster than exports expand. Only the development of new markets, possibly in the developing countries themselves, or of sharply greater access to existing markets in countries belonging to the Organization for Economic Cooperation and Development, can brighten these prospects.

Unfortunately, both international and domestic pressures are pushing in the opposite direction. The global pressures stem primarily from the debt crisis, as developing countries seek to expand exports and restrict imports. To accomplish such an improvement in their trade balance, most developing countries have used a combination of macroeconomic adjustments through exchange rate realignments, more direct controls on imports, and measures to stimulate exports. In the OECD countries, despite much talk of agricultural trade reforms in the context of the Uruguay Round of GATT negotiations, the actual record to date is the opposite. Health concerns over pesticide residues, cyanide in grapes, and hormones in milk and beef, for example, have provided ample excuses to ban the imports of agricultural products that compete with the output of local growers. This anti-import bias for foodstuffs obviously has deep political roots, as the historical evidence presented in Chapter 2 by Lindert shows. The bias is likely to become stronger rather than weaker. New competitive pressures brought about by a massive shift in export orientation on the part of agricultural producers in developing countries, much of it induced by policy conditions imposed by donors, is likely to change longstanding cost relationships. While the United States, for example, preaches the benefits of competitive exchange rates and export-led growth to its aid-recipient countries, it simultaneously finds its own agricultural exports undercut by foreign price competition and in need of direct subsidies to remain competitive. The schizophrenia and doublespeak required of aid officials, the Departments of Agriculture, Treasury, and State, and most commodity organizations in the United States are testimony to the inherent inconsistencies in a more widespread strategy of export-led growth based on agriculture.

A Strategy for Economic Growth

Even granting that agriculture should be a "lead industry" in the growth strategies of most poor countries, policy makers have been presented with two fundamentally different approaches to implementing such strategies. The primary alternative strategy to growth led by exports of agricultural commodities is rural development. As practiced by the World Bank and other major donors, a rural development strategy focuses heavily on raising domestic levels of demand by improving incomes of the rural poor. In principle, much of the additional output produced in rural development projects would be consumed locally. A rising spiral of improved productivity would stimulate production, consumption, and the health and welfare of rural populations. The World Bank's experience, in particular, has demonstrated reasonably satisfactory rates of return on many rural development projects, as Donaldson points out in Chapter 6, but the same experience also highlights a number of important problems with the internal consistency and macroeconomic impact of rural development strategies.

Rural Development versus Agricultural Development. The most crucial question is the relationship—analytically, politically, and bureaucratically—between a country's rural development strategy and its agricultural development strategy. Part of the problem is definitional. There has been a tendency in the profession to accept the World Bank's de facto definition that rural development *projects* must add up to an integrated rural development *strategy.* There has never been any question about the need for an agricultural development strategy to be articulated and implemented by some voice of government. Whether the appropriate visionary and spokesman is the minister of agriculture, the minister of finance, or the president no doubt depends on circumstances, but investments in infrastructure, research, education of farmers, and farm incentives all require an appropriate degree of government involvement.

Is rural development a subset of this involvement, a more encompassing strategic concept, or are agricultural and rural development two separate efforts? Which is responsible for coherence and government leadership? Rural development is broader in some dimensions, especially with its overall concern for rural poverty and the health, education, and general welfare of entire rural communities. But the traditional commodity focus of most agricultural development programs also cuts across rural development projects by providing the research and extension base for raising productivity and by fostering a marketing system that is essential to delivering new inputs and finding profitable outlets for increases in output.

These issues are important because of the macroeconomic significance of the rural economy. If only agricultural sector analysis is capable of providing consistent estimates of the impact of macro polices on the sector, and of the sector on the macro economy, agricultural planning must incorporate planning of rural development projects. If, however, planners are primarily interested in changed levels of poverty and community welfare, rural development strategists must take the lead. Even if administrative integration of rural development activities has been discredited by experience, the issue of strategic integration remains. It is not enough for a few government planners to have an integrated perspective unless they also have the political power and financial resources to guarantee that the various responsible departments implement the necessary pieces in an appropriate fashion.

Roles for the Public and Private Sectors. Whatever the intellectual merits, donors have put rural development strategies in a leading role, thus raising a further set of questions. Because someone must account for the money, who exactly should "do" rural development? It is not an answer to say "the people and their communities." David notes in Chapter 7 that these groups ultimately pay for these projects, although they have virtually no say in their design or implementation. If experience with rural development projects since the 1960s has taught us anything, it is the importance of the economic environment to project success. More to the point, rural communities can be highly resistant to change, because local institutions and power holders find change threatening. Outside resources, personnel, and leadership may well be needed to start this process. The World Bank and other donors have financial and analytical resources, governments have policy levers and personnel in various agencies, and private voluntary organizations often have grass-roots leadership potential. Successful rural development strategies require the integration of these resources into community desires and constraints. Such an integration is not a market activity.

But markets and the private sector obviously play important roles in rural development, if for no other reason than the importance of trade in raising labor productivity in rural areas. And here the pessimism raised by Valdés in the context of export-led growth gains relevance to rural development strategies. Donaldson notes in Chapter 6 that several World Bank–sponsored rural development projects had poor economic returns because prices were low, often because world commodity markets were depressed for the output. Global consistency for price projections will not go away as an issue, even for rural development projects, if higher incomes are to be stimulated by the production of agricultural commodities whose prices are determined in world markets. Both for logical consistency and long-run economic efficiency, rural development projects must be set in the same matrix of trade and development strategies as the rest of the economy.

Stabilizing the Economic Environment. A new orthodoxy of neo-classical policy advice now stresses the efficiency gains from market-directed resource allocations and an outward-oriented trade strategy. "Getting prices right" is usually the central element in this policy advice, and in the new orthodoxy this means free trade.[4] No matter how correct this advice might be for most commodities, for important foodstuffs it is universally ignored. Obviously policy makers are unconvinced that their citizens will be satisfied with a laissez-faire approach to pricing the most important commodity in their market basket—that is, to placing their food security in the hands of impersonal and unstable world markets. In fact, the motive to stabilize food prices extends well beyond the benefits to consumers and includes reducing risks to farmers, encouraging investments and innovation, and helping to stabilize the macro economy. For these reasons, price stabilization to enhance food security is caught in that nebulous world of universal popular acceptance and the widespread skepticism of micro economists about whether the benefits justify the costs.

From the point of view of policy makers and national leaders in developing countries, stabilizing the economic environment, in combination with rising real wages, can be considered an acceptable definition of success for economic policy. But no shortcuts or cheap substitutes can be identified for the policies that create and sustain this combination. Price stabilization requires costly logistical interventions and analytical capacity to manage them; generating higher employment levels for unskilled workers, and thus demand-led increases in their real wages, requires the long-term investments in fiscal discipline, export competitiveness, agricultural infrastructure, and efficient marketing systems that underlay the success stories around the world.

[4]See Lal (1985) for an impassioned argument about the primacy of "right" prices. For an acknowledgment of the importance of agricultural prices but the sharp limits on the usefulness for policy of the border price paradigm in a highly unstable world, see Timmer (1986) and (1989a).

Moving from Policy Analysis to Police Advice

Policy analysis is a feeble tool for moving entire economic systems from their present anti-market, inward-oriented, centrally planned reality toward the more promising end of the spectrum that seems to offer success. But modern economics often makes the task more difficult. As John Lewis has noted in a review of the role of government in national economic development, "Many of our best contemporary economists, whose analyses one reads with awe, live in a kind of two-dimensional space. They find what they do so fascinating—so technically and conceptually exhilarating—that they begin to believe that their findings, in an abstracted slice of reality, provide adequate answers to most problems" (1989, pp. 76–77).

No government or analyst can truly plan for propitious moments when economic reforms suddenly become possible, even inevitable, as the events in Eastern Europe at the end of 1989 made clear. But how should they respond when the opportunity does arise? Precisely because these are major turning points for each society, history has not performed the necessary experiments; every country will have its own unique past, current institutions, and potential avenues of change. At these moments, policy analysts will be hampered, not helped, by ideal visions of a perfect, but two-dimensional, economy in which all of the difficult problems—which markets work and which do not, what objective function firms are using, and the willingness of the population to endure lower standards of living while a select few become visibly rich—have been assumed away. As Lewis notes, there is an "irreducible complexity, even ambiguity, of policy issues" that sharply limits the power of our models to answer the most basic and pressing questions that policy makers ask in moments in crisis (1989, p. 81).

At these times, policy analysts and policy makers inevitably fall back on experience and intuition. Here, the empirical record is crucial. It says that in the long, slow task of national development, agriculture plays a key role. Each chapter in this book argues, for its own set of issues and from the extensive experience of its author, that the state has a role in the various elements of that development process, but so does the private sector. A highly disciplined and effective government can play a relatively larger role; so too can a dynamic and entrepreneurial private sector. Few countries are blessed with both. Finding the right balance between the government and the private sector in the vastly different economic, political, and social environments that characterize the Third World is a difficult task with no simple or general answer. But those societies that find the answer are on the only road to development.

References

Baldwin, Robert E., ed. 1988. *Trade Policy Issues and Empirical Analysis.* Chicago: University of Chicago Press for the National Bureau of Economic Research.

Krugman, Paul R. 1987. "Is Free Trade Passé?" *Journal of Economic Perspectives* 1 (Fall): 131–44.

Lal, Deepak. 1985. *The Poverty of Development Economics.* Cambridge: Harvard University Press.

Lewis, John P. 1989. "Government and National Economic Development." *Daedalus* 118 (Winter): 69–83.

Lipton, Michael. 1977. *Why Poor People Stay Poor: Urban Bias in Developing Countries.* Cambridge: Harvard University Press.

Rodrik, Dani. 1988. "Imperfect Competition, Scale Economies, and Trade Policy in Developing Countries." In Robert E. Baldwin, ed., *Trade Policy Issues and Empirical Analysis*, pp. 109–37. Chicago: University of Chicago Press for the National Bureau of Economic Research.

Timmer, C. Peter. 1986. *Getting Prices Right: The Scope and Limits of Agricultural Price Policy.* Ithaca: Cornell University Press.

———. 1989a. "Food Price Policy: The Rationale for Government Intervention." *Food Policy* 14 (February): 17–27.

———. 1989b. "Indonesia: Transition from Food Importer to Exporter." In Terry Sicular, ed., *Food Price Policy in Asia*, pp. 22–64. Ithaca: Cornell University Press.

Index

Library of Congress Cataloging-in-Publication Data

Agriculture and the state : growth, employment, and poverty in developing countries /
    edited by C. Peter Timmer.
        p. cm. — (Food systems and agrarian change)
    Includes bibliographical references and index.
    ISBN 0–8014–2601–4 (alk. paper). — ISBN 0–8014–9911–9 (pbk. : alk. paper)
        1. Agriculture and state—Developing countries—Congresses.
    2. Agricultural laborers—Developing countries—Congresses.
    3. Rural poor—Developing countries—Congresses. I. Timmer, C.
    Peter. II. Series.
    HD1417.A475   1991
    338. 1'8'091724—dc20   90–27706